Readings for Mathematics: A Humanistic Approach

edited by

G. Joseph Wimbish, Jr.

Oklahoma College of Liberal Arts

Wadsworth Publishing Company, Inc.
Belmont, California

QA
7
W55

Now there are two classes of persons: one class of those who will agree with you and will take your words as a revelation; another class to whom they will be utterly unmeaning, and who will naturally deem them to be idle tales, for they see no sort of profit which is to be obtained from them. And therefore you had better decide at once with which of the two you are proposing to argue. You will very likely say with neither, and that your chief aim in carrying on the argument is your own improvement; at the same time you do not grudge to others any benefit which they may receive.

Plato

Book VII, *Republic*

© 1972 by Wadsworth Publishing Company, Inc., Belmont, California 94002. All rights reserved. No part of this book may be reproduced, stored in a retrieval system or transcribed, in any form or by any means, electronic, mechanical, photocopying, recording or otherwise, without the prior written permission of the publisher.

ISBN-0-534-00154-8

L. C. Cat. Card No. 74-187969

Printed in the United States of America

1 2 3 4 5 6 7 8 9 10 - - - 76 75 74 73 72

Readings for Mathematics: A Humanistic Approach

Table of Contents

Foreword

Preface for the Student   1

Preface for the Teacher   2

Introduction   3

Republic, Book VII   4
Plato

Part One   One "Math," Two "Math," Old "Math," New "Math"   13

Introduction   13

Mathematics as a Creative Art   13
P. R. Halmos

Questions on Halmos   29

The Meaning of Mathematics   30
Morris Kline

Questions on Kline	41
Mathematics as an Element in the History of Thought	42
Alfred North Whitehead	
Questions on Whitehead	46
Part Two Some Historical Facts and Philosophical Beginnings	49
Introduction	49
A Chronology of Mathematical and Nonmathematical Events	50
Republic, Book VII	59
Plato	
Republic, Book X	67
Plato	
Questions on Books VII and X	73
Part Three Potpourri	77
Introduction	77
Critique of Pure Reason	78
Immanuel Kant	
Questions on Kant	81
Through the Looking Glass	82
Lewis Carroll	

Questions on Carroll	*99*
Introduction	*100*
Flatland	*100*
A. Square	
Questions on Flatland	*118*
Part Four What if_____?	*121*
Introduction	*121*
How Mathematicians Develop a Branch of Pure Mathematics	*122*
Harriet F. Montague and Mabel D. Montgomery	
Group Theory and the Postulational Method	*132*
Carl H. Denbow and Victor Goedicke	
Part Five Logic and Proof	*145*
Introduction	*145*
Paradox	*145*
W. V. Quine	
Questions on Paradox	*159*
Symbolic Logic	*160*
John E. Pfeiffer	
Questions for Symbolic Logic	*168*

Part Six The Image of Computation 171

Introduction 171

From: IBM Instructor's Guide 172

Foreword

Many perennial questions arise in conjunction with liberal education. One of the foremost seems to concern just what a liberally educated person is. Unfortunately, no answer seems completely satisfactory, in the sense that no list of characteristics could ever be exhaustive. Indeed, I suspect that if such a list did exist, no one could be found to fit it. Nonetheless, I have my own nonexhaustive criteria for a liberally educated person. One of these is the ability to read with some understanding expository material concerning subject matter areas that are traditionally considered technically difficult.

Historically, mathematics has been an integral part of a liberal education. Too often this part has emphasized the "practical" mechanical skills of solving equations, proving geometric theorems, and making arithmetic computations. Important as these activities are, mathematicians as mathematicians do have other interests. I feel that the liberal arts public should be made aware of this fact.

Often the comment is made that a good course in algebra, trigonometry, or adding fractions would be of more practical value than a course in some more "abstract" mathematical field. What is practical? What is practical to a lathe operator may not be practical to an executive who must communicate with a staff containing many types of technicians. Neither will traditional college algebra be practical to a social or behavioral scientist trying to establish relational patterns between individuals or groups. There are mathematical structures that will aid him if he knows where to find readable expository material.

Mathematics, like every academic discipline, is subject to fads and controversies. Many mathematical controversies are similar to those in other fields. Each of us can learn from other disciplines how to deal with difficulties within our own area of competence—but only if we learn the language of those other disciplines.

Preface for the Student

"Please don't read the
Preface for the Teacher."

Landau
Foundations of Analysis

I have found that many students who face a mathematics course not of their choosing do so with loquacious distaste. This is particularly true when the course seems fairly unrelated to their major interest area. Often, they (rightly) suspect that it will be much the same as high school courses that either bored them or forced them to spend an inordinate amount of time in arduous study.

This volume is an attempt on a small scale to introduce you to a new world of mathematics that may be called expository mathematics. At first, you will probably object to the material and the approach used. You will probably feel that this is not mathematics, especially since it has few numbers in it. But even though the contents often bear little resemblance to what has been foisted onto you as mathematics, this anthology reflects a large segment of mathematical thinking as done by professionals.

In reading this book, students (and new teachers) have found the following suggestions useful. First, read each article through hastily, paying absolutely no attention to details. Try only to catch the flavor, though many times you will feel like Alice after reading "Jabberwocky" (see page 84 of this volume). Then read the article in more detail with pen and paper close at hand; use them freely as follows: (1) read until you find a passage that you either don't understand, strongly agree with, or strongly disagree with; (2) write it down; (3) if you don't understand the passage, ask someone about it; (4) if you agree or disagree with it, explain why in a few sentences; (5) go to statement (1) above. When you have read several articles, begin to compare them; find their similarities and differences.

At the end of each article are self-study questions. These should be used only if you become desperate. They represent some things I felt were important. I am sure that you and your instructor will be able to find many more points of importance. Finally, please talk about these readings outside of class with your classmates and people not taking the course. To further aid your understanding, try to defend a point with which you agree, then one with which you disagree.

Preface for the Teacher

The idea of expository material about mathematical subjects is not a new one. An abundance of material has long been available for those who sought it. Of course, abundance does not of itself imply availability. Moreover, many times that which is readily obtainable is not immediately understandable or usable by the lay reader. Thus, several of us—students as well as teachers—felt that a set of introductory readings should be collected along a certain thematic line for use and study by *all* types of beginning college students, mathematics and nonmathematics majors alike. The present collection is meant to supply some of that material. No such collection could hope to furnish adequately all things needed or wanted by every teacher and student. Indeed, some unwanted material will probably be included. Each person who has considered this problem has definite and differing opinions as to what an anthology of this nature should contain. On the other hand, many worthwhile articles have not been included simply because of permission difficulties and other technical problems. Hence, if your favorite expository article was not included, please try to remember that there were boundary conditions operative.

The course as we teach it at Oklahoma College of Liberal Arts lasts fifteen weeks and meets four times per week. There are two lecture sessions per week coupled with two group discussion periods. These are sometimes group dynamic sessions; at other times they take the form of dialogues between students and leader. In the latter case, the leader attempts to be as Socratic as possible, drawing out the student and attempting to cause him of himself to become more favorably inclined toward mathematics and modes of mathematical thinking. This assumes, rightly or wrongly, that most students who take a liberal arts mathematics course do so under duress.

These readings are used primarily in the "discussion" classes as a supplement to the lectures. They acquaint the student with a wide range of expository literature of varying difficulty.

I have mentioned the lectures. These have provided the basis for the text, published under separate cover, that accompanies these readings.

For further hints on how you might approach this material, please read the Preface for the Student. I hope you enjoy reading and using this anthology as much as we at O.C.L.A. have.

Introduction

As an introduction to this anthology, I have chosen a part of Book VII of Plato's *Republic*. It is commonly called "The Allegory of the Cave." Concerning this segment and Plato's writing in general, some comments are in order.

First, Book VII divides naturally into two segments: (1) the "Allegory" part and (2) everything else. I have accordingly split Book VII into these parts. In so doing, however, I am not denying the obvious continuity and coherence of the dialogue as a unit.

Second, as an aid to understanding, make sure you know what an allegory is. Try to distinguish among the ideas of allegory, parable, simile, and metaphor.

Since more of Plato's writings are included later, I shall give a more complete aid to understanding at that time. However, you should note that Plato clearly would like us to be extremely careful in distinguishing appearance from reality. As an obvious application, one perhaps should not conclude that one knows what mathematics is. There is danger in basing one's conclusions solely on the experiences one has had with prior mathematical education (possibly miseducation). Mathematical shadows are sometimes difficult to tell from reality.

Before you read further, on a convenient page of your notebook, please answer the following questions:

1. What do you think mathematics is (are)?
2. What do mathematicians do and why do they do it?
3. Where would you look for a mathematician if you thought you needed one?

Keep these answers; periodically throughout the course of your reading, re-answer them and note changes in perspective, if any.

Republic, Book VII

Plato

And now, I said, let me show in a figure how far our nature is enlightened or unenlightened:—Behold! human beings housed in an underground cave, which has a long entrance open towards the light and as wide as the interior of the cave; here they have been from their childhood, and have their legs and necks chained, so that they cannot move and can only see before them, being prevented by the chains from turning round their heads. Above and behind them a fire is blazing at a distance, and between the fire and the prisoners there is a raised way; and you will see, if you look, a low wall built along the way, like the screen which marionette players have in front of them, over which they show the puppets.

I see.

And do you see, I said, men passing along the wall carrying all sorts of vessels, and statues and figures of animals made of wood and stone and various materials, which appear over the wall? While carrying their burdens, some of them, as you would expect, are talking, others silent.

You have shown me a strange image, and they are strange prisoners.

Like ourselves, I replied; for in the first place do you think they have seen anything of themselves, and of one another, except the shadows which the fire throws on the opposite wall of the cave?

How could they do so, he asked, if throughout their lives they were never allowed to move their heads?

From *The Dialogues of Plato*, trans. Benjamin Jowett, 4th ed., 1953, Vol. II. Reprinted by permission of the Clarendon Press, Oxford.

And of the objects which are being carried in like manner they would only see the shadows?

Yes, he said.

And if they were able to converse with one another, would they not suppose that the things they saw were the real things?*

Very true.

And suppose further that the prison had an echo which came from the other side, would they not be sure to fancy when one of the passers-by spoke that the voice which they heard came from the passing shadow?

No question, he replied.

To them, I said, the truth would be literally nothing but the shadows of the images.

That is certain.

And now look again, and see in what manner they would be released from their bonds, and cured of their error, whether the process would naturally be as follows. At first, when any of them is liberated and compelled suddenly to stand up and turn his neck round and walk and look towards the light, he will suffer sharp pains; the glare will distress him, and he will be unable to see the realities of which in his former state he had seen the shadows; and then conceive someone saying to him that what he saw before was an illusion, but that now, when he is approaching nearer to being and his eye is turned towards more real existence, he has a clearer vision,—what will be his reply? And you may further imagine that his instructor is pointing to the objects as they pass and requiring him to name them,—will he not be perplexed? Will he not fancy that the shadows which he formerly saw are truer than the objects which are now shown to him?

Far truer.

And if he is compelled to look straight at the light, will he not have a pain in his eyes which will make him turn away to take refuge in the objects of vision which he can see, and which he will conceive to be in reality clearer than the things which are now being shown to him?

True, he said.

And suppose once more, that he is reluctantly dragged up that steep and rugged ascent, and held fast until he is forced into the presence of the sun himself, is he not likely to be pained and irritated? When he approaches the light his eyes will be dazzled, and he will not be able to see anything at all of what are now called realities.

Not all in a moment, he said.

---

*[Text uncertain: perhaps "that they would apply the name *real* to the things which they saw."]

He will require to grow accustomed to the sight of the upper world. And first he will see the shadows best, next the reflections of men and other objects in the water, and then the objects themselves; and, when he turned to the heavenly bodies and the heaven itself, he would find it easier to gaze upon the light of the moon and the stars at night than to see the sun or the light of the sun by day?

Certainly.

Last of all he will be able to see the sun, not turning aside to the illusory reflections of him in the water, but gazing directly at him in his own proper place, and contemplating him as he is.

Certainly.

He will then proceed to argue that this is he who gives the seasons and the years, and is the guardian of all that is in the visible world, and in a certain way the cause of all things which he and his fellows have been accustomed to behold?

Clearly, he said, he would arrive at this conclusion after what he had seen.

And when he remembered his old habitation, and the wisdom of the cave and his fellow-prisoners, do you not suppose that he would felicitate himself on the change, and pity them?

Certainly, he would.

And if they were in the habit of conferring honours among themselves on those who were quickest to observe the passing shadows and to remark which of them went before and which followed after and which were together, and who were best able from these observations to divine the future, do you think that he would be eager for such honours and glories, or envy those who attained honour and sovereignty among those men? Would he not say with Homer, "Better to be a serf, labouring for a landless master," and to endure anything, rather than think as they do and live after their manner?

Yes, he said, I think that he would consent to suffer anything rather than live in this miserable manner.

Imagine once more, I said, such a one coming down suddenly out of the sunlight, and being replaced in his old seat; would he not be certain to have his eyes full of darkness?

To be sure, he said.

And if there were a contest, and he had to compete in measuring the shadows with the prisoners who had never moved out of the cave, while his sight was still weak, and before his eyes had become steady (and the time which would be needed to acquire this new habit of sight might be very considerable), would he not make himself ridiculous? Men would say of him that he had returned from the place above with his eyes ruined; and that it was better not even to think of ascending; and if anyone tried to loose another and lead him up to the light, let them only catch the offender, and they would put him to death.

No question, he said.

This entire allegory, I said, you may now append, dear Glaucon, to the previous argument; the prison-house is the world of sight, the light of the fire is the power of the sun, and you will not misapprehend me if you interpret the journey upwards to be the ascent of the soul into the intellectual world according to my surmise, which, at your desire, I have expressed—whether rightly or wrongly God knows. But, whether true or false, my opinion is that in the world of knowledge the Idea of good appears last of all, and is seen only with an effort; although, when seen, it is inferred to be the universal author of all things beautiful and right, parent of light and of the lord of light in the visible world, and the immediate and supreme source of reason and truth in the intellectual; and that this is the power upon which he who would act rationally either in public or private life must have his eye fixed.

I agree, he said, as far as I am able to understand you.

Moreover, I said, you must agree once more, and not wonder that those who attain to this vision are unwilling to take any part in human affairs; for their souls are ever hastening into the upper world where they desire to dwell; which desire of theirs is very natural, if our allegory may be trusted.

Yes, very natural.

And is there anything surprising in one who passes from divine contemplations to the evil state of man, appearing grotesque and ridiculous; if, while his eyes are blinking and before he has become accustomed to the surrounding darkness, he is compelled to fight in courts of law, or in other places, about the images or the shadows of images of justice, and must strive against some rival about opinions of these things which are entertained by men who have never yet seen the true justice?

Anything but surprising, he replied.

Anyone who has common sense will remember that the bewilderments of the eyes are of two kinds and arise from two causes, either from coming out of the light or from going into the light, and, judging that the soul may be affected in the same way, will not give way to foolish laughter when he sees anyone whose vision is perplexed and weak; he will first ask whether that soul of man has come out of the brighter life and is unable to see because unaccustomed to the dark, or having turned from darkness to the day is dazzled by excess of light. And he will count the one happy in his condition and state of being, and he will pity the other; or, if he have a mind to laugh at the soul which comes from below into the light, this laughter will not be quite so laughable as that which greets the soul which returns from above out of the light into the cave.

That, he said, is a very just distinction.

But then, if I am right, certain professors of education must be wrong when they say that they can put a knowledge into the soul which was not there before, like sight into blind eyes.

They undoubtedly say this, he replied.

Whereas our argument shows that the power and capacity of learning exists in the soul already; and that just as if it were not possible to turn the eye from darkness to light without the whole body, so too the instrument of knowledge can only by the movement of the whole soul be turned from the world of becoming to that of being, and learn by degrees to endure the sight of being, and of the brightest and best of being, or in other words, of the good.

Very true.

And must there not be some art which will show how the conversion can be effected in the easiest and quickest manner; an art which will not implant the faculty of sight, for that exists already, but will set it straight when it has been turned in the wrong direction, and is looking away from the truth?

Yes, he said, such an art may be presumed.

And whereas the other so-called virtues of the soul seem to be akin to bodily qualities, for even when they are not originally innate they can be implanted later by habit and exercise, the virtue of wisdom more than anything else contains a divine element which never loses its power, and by this conversion is rendered useful and profitable; or, by conversion of another sort, hurtful and useless. Did you never observe the narrow intelligence flashing from the keen eye of a clever rogue—how eager he is, how clearly his paltry soul sees the way to his end; he is the reverse of blind, but his keen eye-sight is forced into the service of evil, and he is mischievous in proportion to his cleverness?

Very true, he said.

But what if such natures had been gradually stripped, beginning in childhood, of the leaden weights which sink them in the sea of Becoming, and which, fastened upon the soul through gluttonous indulgence in eating and other such pleasures, forcibly turn its vision downwards—if, I say, they had been released from these impediments and turned in the opposite direction, the very same faculty in them would have seen the truth as keenly as they see what their eyes are turned to now.

Very likely.

Yes, I said; and there is another thing which is likely, or rather a necessary inference from what has preceded, that neither the uneducated and uninformed of the truth, nor yet those who are suffered to prolong their education without end, will be able ministers of State; not the former, because they have no single aim of duty which is the rule of all their actions, private as well as public; nor the latter, because they will not act at all except upon compulsion, fancying that they are already dwelling apart in the islands of the blest.

Very true, he replied.

Then, I said, the business of us who are the founders of the State will be to compel the best minds to attain that knowledge which we have already shown to be the greatest of all, namely, the vision of the good; they must make the ascent which we have described; but when they have ascended and seen enough we must not allow them to do as they do now.

What do you mean?

They are permitted to remain in the upper world, refusing to descend again among the prisoners in the cave, and partake of their labours and honours, whether they are worth having or not.

But is not this unjust? he said; ought we to give them a worse life, when they might have a better?

You have again forgotten, my friend, I said, the intention of our law, which does not aim at making any one class in the State happy above the rest; it seeks rather to spread happiness over the whole State, and to hold the citizens together by persuasion and necessity, making each share with others any benefit which he can confer upon the State; and the law aims at producing such citizens, not that they may be left to please themselves, but that they may serve in binding the State together.

True, he said, I had forgotten.

Part One

One "Math," Two "Math," Old "Math," New "Math"

Introduction

The first three articles in this section were written by twentieth-century mathematicians. The articles were chosen for several reasons. Each of the three authors, Halmos, Kline, and Whitehead, is articulate both in technical and expositorial mathematical writing. All three are concerned with establishing lines of communication among the various subcultures of our society. Their beliefs concerning mathematics and mathematicians should be very apparent from their articles.

These men, however, have many differing opinions about mathematics, its nature, its nurture, and its use. The companion text maintains a threefold approach to human beings. This approach declares the importance of language, thought, and phenomenological systems in the structuring of a man's being. Mathematics is a human activity, and so must contribute to and be influenced by each of the above manifolds of human existence. Look for these items as you read.

These three men have written much good expository material. I encourage you to find and read it.

Mathematics as a Creative Art

P. R. Halmos

Do you know any mathematicians—and, if you do, do you know anything about what they do with their time? Most people don't. When I get into conversation

From "Mathematics As a Creative Art" by P. R. Halmos in *American Scientist*, Vol. 56, No. 4, pp. 375–389, 1968. Reprinted by permission of the publisher.

with the man next to me in a plane, and he tells me that he is something respectable like a doctor, lawyer, merchant, or dean, I am tempted to say that I am in roofing and siding. If I tell him that I am a mathematician, his most likely reply will be that he himself could never balance his check book, and it must be fun to be a whiz at math. If my neighbor is an astronomer, a biologist, a chemist, or any other kind of natural or social scientist, I am, if anything, worse off—this man *"thinks"* he knows what a mathematician is, and he is probably wrong. He thinks that I spend my time (or should) converting different orders of magnitude, comparing binomial coefficients and powers of 2, or solving equations involving rates of reactions.

C. P. Snow points to and deplores the existence of two cultures; he worries about the physicist whose idea of modern literature is Dickens, and he chides the poet who cannot state the second law of thermodynamics. Mathematicians, in converse with well-meaning, intelligent, and educated laymen (do you mind if I refer to all non-mathematicians as laymen?) are much worse off than physicists in converse with poets. It saddens me that educated people don't even know that my subject exists. There is something that they call mathematics, but they neither know how the professionals use that word, nor can they conceive why anybody should do it. It is, to be sure, possible that an intelligent and otherwise educated person doesn't know that egyptology exists, or haematology, but all you have to tell him is that it does, and he will immediately understand in a rough general way why it should and he will have some empathy with the scholar of the subject who finds it interesting.

Usually when a mathematician lectures, he is a missionary. Whether he is talking over a cup of coffee with a collaborator, lecturing to a graduate class of specialists, teaching a reluctant group of freshman engineers, or addressing a general audience of laymen—he is still preaching and seeking to make converts. He will state theorems and he will discuss proofs and he will hope that when he is done his audience will know more mathematics than they did before. My aim today is different—I am not here to proselyte but to enlighten—I seek not converts but friends. I do not want to teach you what mathematics is, but only *"that"* it is.

I call my subject mathematics—that's what all my colleagues call it, all over the the world—and there, quite possibly, is the beginning of confusion. The word covers two disciplines—many more, in reality, but two, at least two, in the same sense in which Snow speaks of two cultures. In order to have some words with which to refer to the ideas I want to discuss, I offer two temporary and ad hoc neologisms. Mathematics, as the word is customarily used, consists of at least two distinct subjects, and I propose to call them *"mathology"* and *"mathophysics."* Roughly speaking, mathology is what is usually called pure mathematics, and mathophysics is called applied mathematics, but the qualifiers are not emotionally strong enough to disguise that they qualify the same noun. If the concatenation of syllables I chose here reminds you of other words, no great harm will be done; the rhymes alluded to are not completely accidental. I originally planned to entitle this lecture something like "Mathematics is an art," or "Mathematics is not a science," or "Mathematics is useless," but the more I thought about it the more I realized that I mean that "Mathology is an art," "Mathology is not a science," and "Mathology is useless." When I am through, I hope you will recognize that most of you have known about mathophysics before, only you were probably calling it mathematics; I hope that all of you will recognize the distinction between mathology and mathophysics; and I hope that some of you will be ready to embrace, or at least applaud, or at the very least, recognize mathology as a respectable human endeavor.

In the course of the lecture I'll have to use many analogies (literature, chess, painting), each imperfect by itself, but I hope that in their totality they will serve to delineate what I want delineated. Sometimes in the interest of economy of time, and sometimes doubtless unintentionally, I'll exaggerate. When I'm done, I'll be glad to rescind anything that was inaccurate or that gave offense in any other way.

What Mathematicians Do

As a first step toward telling you what mathematicians do, let me tell you some of the things they do not do. To begin with, mathematicians have very little to do with numbers. You can no more expect a mathematician to be able to add a column of figures rapidly and correctly than you can expect a painter to draw a straight line or a surgeon to carve a turkey—popular legend attributes such skills to these professions, but popular legend is wrong. There is, to be sure, a part of mathematics called number theory, but even that doesn't deal with numbers in the legendary sense—a number theorist and an adding machine would find very little to talk about. A machine might enjoy proving that $1^3 + 5^3 + 3^3 = 153$, and it might even go on to discover that there are only five positive integers with the property that the equation indicates (1, 370, 371, 407), but most mathematicians couldn't care less; many mathematicians enjoy and respect the theorem that every positive integer is the sum of not more than four squares, whereas the infinity involved in the word "every" would frighten and paralyze any ordinary office machine, and, in any case, that's probably not the sort of thing that the person who relegates mathematics to numbers had in mind.

Not even those romantic objects of latter day science fiction, the giant brains, the computing machines that run our lives these days—not even they are of interest to the mathematician as such. Some mathematicians are interested in the logical problems involved in the reduction of difficult questions to the sort of moronic baby talk that machines understand; the logical design of computing machines is definitely mathematics. Their construction is not, that's engineering, and their product, be it a payroll, a batch of sorted mail, or a supersonic plane, is of no mathematical interest or value.

Mathematics is not numbers or machines; it is also not the determination of the heights of mountains by trigonometry, or compound interest by algebra, or moments of inertia by calculus. Not today it isn't. At one point in history each of those things, and others like them, might have been an important and nontrivial research problem, but once the problem is solved, its repetitive application has as much to do with mathematics as the work of a Western Union messenger boy has to do with Marconi's genius.

There are at least two other things that mathematics isn't; one of them is something it never was, and the other is something it once included and by now has sloughed off. The first is physics. Some laymen confuse mathematics and theoretical physics and speak, for instance, of Einstein as a great mathematician. There is no doubt that Einstein was a great man, but he was no more a great mathematician than he was a great violinist. He used mathematics to find out facts about the universe, and that he successfully used certain parts of differential geometry for that purpose adds a certain piquancy to the appeal of differential geometry. Withal, relativity theory and differential geometry are not the same thing. Einstein, Schrodinger, Heisenberg, Fermi, Wigner, Feynman—great men all, but not mathematicians; some of them, in fact, strongly antimathematical, preach against mathematics, and would regard it as an insult to be called a mathematician.

What once was mathematics remains mathematics always, but it can become so thoroughly worked out, so completely understood, and, in the light of millennia of contributions, with hindsight, so trivial, that mathematicians never again need to or want to spend time on it. The celebrated Greek problems (trisect the angle, square the circle, duplicate the cube) are of this kind, and the irrepressible mathematical amateur to the contrary notwithstanding, mathematicians are no longer trying to solve them. Please understand, it isn't that they have given up. Perhaps you have heard that, according to mathematicians, it is impossible to square a circle, or trisect an angle, and perhaps you have heard or read that, therefore, mathematicians are a pusillanimous chicken-hearted lot, who give up easily, and use their ex-cathedra pronouncements to justify their ignorance. The conclusion may be true, and you may believe it if you like, but the proof is inadequate. The point is a small one but a famous one and one of historical interest: let me digress to discuss it for a moment.

A Short Digression

The problem of trisecting the angle is this: given an angle, construct another one that is just one third as large. The problem is perfectly easy, and several methods for solving it are known. The catch is that the original Greek formulation of the problem is more stringent: it requires a construction that uses ruler and compasses only. Even that can be done, and I could show you a perfectly simple method in one minute and convince you that it works in two more minutes. The real difficulty is that the precise formulation of the problem is more stringent still. The precise formulation demands a construction that uses a ruler and compasses only and, moreover, severely restricts how they are to be used; it prohibits, for instance, marking two points on the ruler and using the marked points in further constructions. It takes some careful legalism (or some moderately pedantic mathematics) to formulate really precisely just what was and what wasn't allowed by the Greek rules. The modern angle trisector either doesn't know those rules, or he knows them but thinks that the idea is to get a close approximation, or he knows the rules and knows that an exact solution is required but lets wish be father to the deed and simply makes a mistake. Frequently his attitude is that of the visitor from outer space to golf. (If all you want is to get that little white ball in that little green hole, why don't you just go and put it there?)

Allow me to add a short digression to the digression. I'd like to remind you that when a mathematician says that something is impossible, he doesn't mean that it is very very difficult, beyond his powers, and probably beyond the powers of all humanity for the foreseeable future. That's what is often meant when someone says it's impossible to travel at the speed of sound five miles above the surface of the earth, or instantaneously to communicate with someone a thousand miles away, or to tamper with the genetic code so as to produce a race of citizens who are simultaneously intelligent and peace-loving. That's what is belittled by the classic business braggadocio (the impossible takes a little longer). The mathematical impossible is different: it is more modest and more secure. The mathematical impossible is the logical impossible. When the mathematician says that it is impossible to find a positive number whose sum with 10 is less than 10, here merely reminds us that

that's what the words mean (positive, sum, 10, less); when he says that it is impossible to trisect every angle by ruler and compasses, he means exactly the same sort of thing, only the number of technical words involved is large enough and the argument that strings them together is long enough that they fill a book, not just a line.

The Start of Mathematics

No one knows when and where mathematics got started, or how, but it seems reasonable to guess that it emerged from the same primitive physical observations (counting, measuring) with which we all begin our own mathematical insight (ontogeny recapitulates phylogeny). It was probably so in the beginning, and it is true still, that many mathematical ideas originate not from pure thought but from material necessity; many, but probably not all. Almost as soon as a human being finds it necessary to count his sheep (or sooner?) he begins to wonder about numbers and shapes and motions and arrangements—curiosity about such things seems to be as necessary to the human spirit as curiosity about earth, water, fire, and air, and curiosity—sheer pure intellectual curiosity—about stars and about life. Numbers and shapes and motions and arrangements, and also thoughts and their order, and concepts such as "property" and "relation"—all such things are the raw material of mathematics. The technical but basic mathematical concept of "group" is the best humanity can do to understand the intuitive concept of "symmetry" and the people who study topological spaces, and ergodic paths, and oriented graphs are making precise our crude and vague feelings about shapes, and motions, and arrangements.

Why do mathematicians study such things, and why should they? What, in other words, motivates the individual mathematician, and why does society encourage his efforts, at least to the extent of providing him with the training and subsequently the livelihood that, in turn, give him the time he needs to think? There are two answers to each of the two questions: because mathematics is practical and because mathematics is an art. The already existing mathematics grows and the number of people who think about it keeps doubling over and over again, more new concepts need explication, more new logical interrelations cry out for study, and understanding, and simplification, and more and more the tree of mathematics bears elaborate and gaudy flowers that are, to many beholders, worth more than the roots from which it all comes and the causes that brought it all into existence.

Mathematics Today

Mathematics is very much alive today. There are more than a thousand journals that publish mathematical articles; about 15,000 to 20,000 mathematical articles are printed every year. The mathematical achievements of the last 100 years are greater in quantity and in quality than those of all previous history. Difficult mathematical problems, which stumped Hilbert, Cantor, or Poincaré, are being solved, explained, and generalized by beardless (and bearded) youths in Berkeley and in Odessa.

Mathematicians sometimes classify themselves and each other as either problem-solvers or theory-creators. The problem-solvers answer yes-or-no questions and discuss the vital special cases and concrete examples that are the flesh and blood of mathematics; the theory creators fit the results into a framework, illuminate it all, and point it in a definite direction—they provide the skeleton and the soul of mathematics. One and the same human being can be both a problem-solver and a theory-creator, but, usually, he is mainly one or the other. The problem-solvers make geometric constructions, the theory-creators discuss the foundations of Euclidean geometry; the problem-solvers find out what makes switching diagrams tick, the theory-creators prove representation theorems for Boolean algebras. In both kinds of mathematics and in all fields of mathematics the progress in one generation is breath-taking. No one can call himself a mathematician nowadays who doesn't have at least a vague idea of homological algebra, differential topology, and functional analysis, and every mathematician is probably somewhat of an expert on at least one of these subjects— and yet when I studied mathematics in the 1930's none of those phrases had been invented, and the subjects they describe existed in seminal forms only.

Mathematics is abstract thought, mathematics is pure logic, mathematics is creative art. All these statements are wrong, but they are all a little right, and they are all nearer the mark than "mathematics is numbers" or "mathematics is geometric shapes." For the professional pure mathematician, mathematics is the logical dovetailing of a carefully selected sparse set of assumptions with their surprising conclusions via a conceptually elegant proof. Simplicity, intricacy, and above all, logical analysis are the hallmark of mathematics.

The mathematician is interested in extreme cases—in this respect he is like the industrial experimenter who breaks lightbulbs, tears shirts, and bounces cars on ruts. How widely does a reasoning apply, he wants to know, and what happens when it doesn't? What happens when you weaken one of the assumptions, or under what conditions can you strengthen one of the conclusions? It is the perpetual asking of such questions that makes for broader understanding, better technique, and greater elasticity for future problems.

Mathematics—this may surprise you or shock you some—is never deductive in its creation. The mathematician at work makes vague guesses, visualizes broad generalizations, and jumps to unwarranted conclusions. He arranges and rearranges his ideas, and he becomes convinced of their truth long before he can write down a logical proof. The conviction is not likely to come early—it usually comes after many attempts, many failures, many discouragements, many false starts. It often happens that months of work result in the proof that the method of attack they were based on cannot possibly work, and the process of guessing, visualizing, and conclusion-jumping begins again. A reformulation is needed—and—and this too may surprise—more experimental work is needed. To be sure, by "experimental work" I do not mean test tubes and cyclotrons. I mean thought-experiments. When a mathematician wants to prove a theorem about an infinite-dimensional Hilbert space, he examines its finite-dimensional analogue, he looks in detail at the 2- and 3-dimensional cases, he often tries out a particular numerical case, and he hopes that he will gain thereby an insight that pure definition-juggling has not yielded. The deductive stage, writing the result down, and writing down its rigorous proof are relatively trivial once the real insight arrives; it is more like the draftsman's work, not the architect's.

The Mathematical Fraternity

The mathematical fraternity is a little like a self-perpetuating priesthood. The mathematicians of today train the mathematicians of tomorrow and, in effect, decide whom to admit to the priesthood. Most people do not find it easy to join—mathematical talent and genius are apparently exactly as rare as talent and genius in painting and music—but anyone can join, everyone is welcome. The rules are nowhere explicitly formulated, but they are intuitively felt by everyone in the profession. Mistakes are forgiven and so is obscure exposition—the indispensable requisite is mathematical insight. Sloppy thinking, verbosity without content, and polemic have no role, and—this to me is one of the most wonderful aspects of mathematics—they are much easier to spot than in the non-mathematical fields of human endeavor (much easier than, for instance, in literature among the arts, in art criticism among the humanities, and in your favorite abomination among the social sciences).

Although most of mathematical creation is done by one man at a desk, at a blackboard, or taking a walk, or, sometimes, by two men in conversation, mathematics is nevertheless a sociable science. The creator needs stimulation while he is creating and he needs an audience after he has created. Mathematics is a sociable science in the sense that I don't think it can be done by one man on a desert island (except for a very short time), but it is not a mob science, it is not a team science. A theorem is not a pyramid; inspiration has never been known to descend on a committee. A great theorem can no more be obtained by a "project" approach than a great painting; I don't think a team of little Gausses could have obtained the theorem about regular polygons under the leadership of a rear admiral anymore than a team of little Shakespeares could have written *Hamlet* under such conditions.

A Tiny and Trivial Mathematical Problem

I have been trying to give you a description of what mathematics is and how mathematicians do it, in broad general terms, and I wouldn't blame you if you had been finding it thoroughly unsatisfactory. I feel a little as if I had been describing snow to a Fiji Islander. If I told him snow was white like an egg, wet like mud, and cold like a mountain waterfall, would he then understand what it's like to ski in the Alps? To show him a spoonful of scrapings from the just defrosted refrigerator of His Excellency the Governor is not much more satisfactory—but it is a little. Let me, therefore, conclude this particular tack by mentioning a tiny and trivial mathematical problem and describing its solution—possibly you'll then get (if you don't already have) a little feeling for what attracts and amuses mathematicians and what is the nature of the inspiration I have been talking about.

Imagine a society of 1025 tennis players. The mathematically minded ones among you, if you haven't already heard about this famous problem, have immediately been alerted by the number. It is known to anyone who ever kept on doubling something, anything, that 1024 is 2^{10}. All cognoscenti know, therefore, that the presence in the statement of a problem of a number like $1 + 2^{10}$ is bound to be a strong hint to its solution; the chances are, and this can be guessed even before the statement of the problem is complete, that the solution will depend on doubling—or halving—something ten times. The more knowledgeable cognoscenti will also admit the possibility that the number is not a hint but a trap. Imagine then that the tennis players are about to conduct a gigantic tournament, in the following manner. They draw lots to pair off as far as they can, the odd man sits out the first round, and the paired players play their matches. In the second round only the winners of the first round participate, and the whilom odd man. The procedure is the same for the second round as for the first—pair off and play at random, with the new odd man (if any) waiting it out. The rules demand that this procedure be continued, over and over again, until the champion of the society is selected. The champion, in this sense, didn't exactly beat everyone else, but he can say, of each of his fellow players, that he beat some one, who beat some one, ..., who beat some one, who beat that player. The question is: how many matches were played altogether, in all the rounds of the whole tournament?

There are several ways of attacking the problem, and even the most naive one works. According to it, the first round has 512 matches (since 1025 is odd and 512 is a half of 1024), the second has 256 (since 512 winners in the first round, together with the odd man of that round, make 513, which is odd again, and 256 is a half of 512), etc. The "etcetera" yields, after 512 and 256, the numbers 128, 64, 32, 61, 8, 4, 2, 1, and 1 (the very last round, consisting of only one match, is the only one where there is no odd man), and all that is necessary is to add them up. That's a simple job that pencil and paper can accomplish in a few seconds; the answer (and hence the solution of the problem is 1024.

The mathematical wiseacre would proceed a little differently. He would quickly recognize, as advertised, that the problem has to do with repeated halvings, so that the numbers to be added up are the successive powers of 2, from the ninth down to the first,—no, from ninth down to the zeroth!—together with the last 1 caused by the obviously malicious attempt of the problem-setter to confuse the problem-solver by using 1025 instead of 1024. The wiseacre would then proudly exhibit his knowledge of the formula for the sum of a geometric progression, he would therefore know (without addition) that the sum of 512, 256, ..., 8, 4, 2, and 1 is 1023, and he would then add the odd 1 to get the same total of 1024.

The trouble with the wiseacre's solution is that it's much too special. If the number of tennis players had been 1000 instead of 1025, the wiseacre would be no better off than the naive layman. The wiseacre's solution works, but it is as free of inspiration as the layman's. It is shorter but it is still, in the mathematician's contemptuous word, computational.

The problem has also an inspired solution that requires no computation, no formulas, no numbers—just pure thought. Reason like this: each match has a winner and a loser. A loser cannot participate in any later rounds; every one in the society, except only the champion, loses exactly one match. There are, therefore, exactly as many matches as there are losers, and, consequently, the number of matches is exactly one less than the membership of the society. If the number had been 1000, the answer would be 999, and, obviously, the present pure thought method gives the answer, with no computation, for every possible number of players.

That's it: that's what I offer as a microcosmic example of a pretty piece of mathematics. The example is bad because, after all my warning that mathematicians are interested in other things than counting, it deals with counting; it's bad because it does not, cannot, exhibit any of the conceptual power and intellectual technique of nontrivial mathematics; and it's bad because it illustrates applied mathematics (that is, mathematics as applied to a "real life" problem) more than it illustrates pure mathematics (that is, the distilled form of a question about the logical interrelations of concepts—concepts, not tennis players, and tournaments, and matches). For an example, for a parable, it does pretty well nevertheless; if your imagination is good enough mentally to reconstruct the ocean from a drop of water, then you can reconstruct mathematics from the problem of the tennis players.

Mathology vs. Mathophysics

I've been describing mathematics, but, the truth to tell, I've had mathology (pure) in mind, more than mathophysics (applied). For some reason the practitioners of mathophysics tend to minimize the differences between the two subjects and the others, the mathologists, tend to emphasize them. You've long ago found me out, I am sure. Every mathematician is in one camp or the other (well, almost every—a few are in both camps), and I am a mathologist by birth and training. But in a report, such as this one, I must try not to exaggerate my prejudices, so I'll begin by saying that the similarities between mathology and mathophysics are great indeed. It is a historical fact that ultimately all mathematics comes to us, is suggested to us, by the physical universe; in that sense all mathematics is applied. It is, I believe, a psychological fact that even the purest of the pure among us is just a wee bit thrilled when his thoughts make a new and unexpected contact with the non-mathematical universe. The kind of talent required to be good in mathology is intimately related to the kind that mathophysics demands. The articles that mathophysicists write are frequently indistinguishable from those of their mathological colleagues.

As I see it, the main difference between mathophysics and mathology is the *purpose* of the intellectual curiosity that motivated the work—or, perhaps, it would be more accurate to say that it is the *kind* of intellectual curiosity that is relevant. Let me ask you a peculiar but definitely mathematical question. Can you load a pair of dice so that all possible rolls—better: all possible sums that can show on one roll, all the numbers between 2 and 12 inclusive—are equally likely? The question is a legitimate piece of mathematics; the answer to it is known, and it is not trivial. I mention it here so that you may perform a quick do-it-yourself psychoanalysis on yourself. When I asked the question, did you think of homogeneous and nonhomogeneous distributions of mass spread around in curious ways through two cubes, or did you think of sums of products of twelve numbers (the twice six probabilities associated with the twice six faces of the two dice)? If the former, you are a crypto-mathophysicist, if the latter you are a potential mathologist.

How do you choose your research problem, and what about it attracts you? Do you want to know about nature or about logic? Do you prefer concrete facts to abstract relations? If it's nature you want to study, if the concrete has the greater appeal, then you are a mathophysicist. In mathophysics the question always comes from the outside, from the "real world," and the satisfaction the scientist gets from the solution comes, to a large extent, from the light it throws on *facts*.

Surely no one can object to mathophysics or think less of it for that; and yet many do. I did not mean to identify "concrete" with "practical" and thereby belittle it, and equally I did not mean to identify "abstract" with "useless." (That $2^{11213} - 1$ is a prime is a concrete fact, but surely a useless one; that $E = mc^2$ is an abstract relation but unfortunately a practical one.) Nevertheless, such identifications—applied–concrete–practical–crude and pure–abstract–pedantic–useless—are quite common in both camps. To the applied mathematician the antonym of "applied" is "worthless," and to the pure mathematician the antonym of "pure" is "dirty."

History doesn't help the confusion. Historically, pure and applied mathematics (mathology and mathophysics) have been much closer together than they are today. By now the very terminology (pure mathematics versus applied mathematics) makes for semantic confusion: it implies identity with small differences, instead of diversity with important connections.

From the difference in purposes follows a difference in tastes and hence of value judgments. The mathophysicist wants to know the facts, and he has, sometimes at any rate, no patience for the hair-splitting pedantry of the mathologist's rigor (which he derides as rigor mortis). The mathologist wants to understand the ideas, and he places great value on the aesthetic aspects of the understanding and the way that understanding is arrived at; he uses words such as "elegant" to describe a proof. In motivation, in purpose, frequently in method, and almost always in taste, the mathophysicist and the mathologist differ.

When I tell you that I am a mathologist, I am not trying to defend useless knowledge, or convert you to the view that it's the best kind. I would, however, be less than honest with you if I didn't tell you that I believe that. I like the idea of things being done for their own sake. I like it in music, I like it in the crafts, and I like it even in medicine. I never quite trust a doctor who says that he chose his profession out of a desire to benefit humanity; I am uncomfortable and skeptical when I hear such things. I much prefer the doctor to say that he became one because he got good grades in high school zoology. I like the subject for its own sake, in medicine as much as in music; and I like it in mathematics.

Let me digress for a moment to a brief and perhaps apocryphal story about David Hilbert, probably the greatest mathematician of both the nineteenth and twentieth centuries. When he was preparing a public address, Hilbert was asked to include a reference to the conflict (even then!) between pure and applied mathematics, in the hope that if anyone could take a step toward resolving it, he could. Obediently, he is said to have begun his address by saying "I was asked to speak about the conflict between pure and applied mathematics. I am glad to do so, because it is, indeed, a lot of nonsense—there should be no conflict, there can be no conflict—there is no conflict—in fact the two have nothing whatsoever to do with one another!"

It is, I think, undeniable that a great part of mathematics was born, and lives in respect and admiration, for no other reason than that it is interesting—it is interesting in itself. The angle trisection of the Greeks, the celebrated four-color map problem, and Gödel's spectacular contribution to mathematical logic are good because they are beautiful, because they are surprising, because we want to know. Don't all of us feel the irresistible pull of the puzzle? Is there really something wrong with saying that mathematics is a glorious creation of the human spirit and deserves to live even in the absence of any practical application?

Mathematics is a Language

Why does mathematics occupy such an isolated position in the intellectual firmament? Why is it good form, for intellectuals, to shudder and announce that they can't bear it, or, at the very least, to giggle and announce that they never could understand it? One reason, perhaps, is that mathematics is a language. Mathematics is a special and subtle language designed to express certain kinds of ideas more briefly, more accurately, and more usefully than ordinary language. I do not mean here that mathematicians, like members of all other professional cliques, use jargon. They do, at times, and they don't most often, but that's a personal phenomenon, not the professional one I am describing. What I do mean by saying that mathematics is a language is sketchily and inadequately illustrated by the difference between the following two sentences. (1) If each of two numbers is multiplied by itself, the difference of the two results is the same as the product of the sum of the two given numbers by their difference. (2) $x^2 - y^2 = (x + y)(x - y)$. (Note: the longer formulation is not only awkward, it is also incomplete.)

One thing that sometimes upsets and repels the layman is the terminology that mathematicians employ. Mathematical words are intended merely as labels, sometimes suggestive, possibly facetious, but always precisely defined; their everyday connotations must be steadfastly ignored. Just as nobody nowadays infers from the name Fitzgerald that its bearer is the illegitimate son of Gerald, a number that is called irrational must not be thought unreasonable; just as a dramatic poem called *The Divine Comedy* is not necessarily funny, a number called imaginary has the same kind of mathematical existence as any other. (Rational, for numbers, refers not to the Latin ratio, in the sense of reason, but to the English "ratio," in the sense of quotient.)

Mathematics is a language. None of us feels insulted when a sinologist uses Chinese phrases, and we are resigned to living without Chinese, or else spending years learning it. Our attitude to mathematics should be the same. It's a language, and it takes years to learn to speak it well. We all speak it a little, just because some of it is in the air all the time, but we speak it with an accent and frequently inaccurately; most of us speak it, say, about as well as one who can only say "Oui, monsieur" and "S'il vous plaît" speaks French. The mathematician sees nothing wrong with this as long as he's not upbraided by the rest of the intellectual community for keeping secrets. It took him a long time to learn his language, and he doesn't look down on the friend who, never having studied it, doesn't speak it. It is however sometimes difficult to keep one's temper with the cocktail party acquaintance who demands that he be taught the language between drinks and who regards failure or refusal to do so as sure signs of stupidity or snobbishness.

Some Analogies

A little feeling for the nature of mathematics and mathematical thinking can be got by the comparison with chess. The analogy, like all analogies, is imperfect, but it is illuminating just the same. The rules for chess are as arbitrary as the axioms of mathematics sometimes seem to be. The game of chess is as abstract as mathematics. (That chess is played with solid pieces, made of wood, or plastic, or glass, is not an intrinsic feature of the game. It can just as well be played with pencil and paper, as mathematics is, or blindfold, as mathematics can.) Chess also has its elaborate technical language, and chess is completely deterministic.

There is also some analogy between mathematics and music. The mathologist feels the need to justify pure mathematics exactly as little as the musician feels the need to justify music. Do practical men, the men who meet payrolls, demand only practical music—soothing jazz to make an assembly line worker turn nuts quicker, or stirring marches to make a soldier kill with more enthusiasm? No, surely none of us believes in that kind of justification; music, and mathematics, are of human value because human beings feel they are.

The analogy with music can be stretched a little further. Before a performer's artistic contribution is judged, it is taken for granted that he hits the right notes, but merely hitting the right notes doesn't make him a musician. We don't get the point of painting if we compliment the nude Maya on being a good likeness, and we don't get the point of a historian's work if all we can say is that he didn't tell lies. Mere accuracy in performance, resemblance in appearance, and truth in storytelling doesn't make good music, painting, history: in the same way, mere logical correctness doesn't make good mathematics.

Goodness, high quality, are judged on grounds more important than validity, but less describable. A good piece of mathematics is connected with much other mathematics, it is new without being silly (think of a "new" western movie in which the names and the costumes are changed, but the plot isn't), and it is deep in an ineffable but inescapable sense—the sense in which Johann Sebastian is deep and Carl Philip Emmanuel is not. The criterion for quality is beauty, intricacy, neatness, elegance, satisfaction, appropriateness—all subjective, but all somehow mysteriously shared by all.

Mathematics resembles literature also, differently from the way it resembles music. The writing and reading of literature are related to the writing and reading of newspapers, advertisements, and road signs the way mathematics is related to practical arithmetic. We all need to read and write and figure for daily life; but literature is more than reading and writing, and mathematics is more than figuring. The literature analogy can be used to help understand the role of teachers and the role of the pure-applied dualism.

Many whose interests are in language, in the structure, in the history, and in the aesthetics of it, earn their bread and butter by teaching the rudiments of language to its future practical users. Similarly many, perhaps most, whose interests are in the mathematics of today, earn their bread and butter by teaching arithmetic, trigonometry, or calculus. This is sound economics: society abstractly and impersonally is willing to subsidize pure language and pure mathematics, but not very far. Let the would-be purist pull his weight by teaching the next generation the applied aspects of his craft; then he is permitted to spend a fraction of his time doing what he prefers. From the point of view of what a good teacher must be, this is good. A teacher must know more than the bare minimum he must teach; he must know more in order to avoid more and more mistakes, to avoid the perpetuation of misunderstanding, to avoid catastrophic educational inefficiency. To keep him alive, to keep him from drying up, his interest in syntax, his burrowing in etymology, or his dabbling in poetry, play a necessary role.

The pure-applied dualism exists in literature, too. The source of literature is human life, but literature is not the life it comes from, and writing with a grim purpose is not literature. Sure there are borderline cases: is Upton Sinclair's "Jungle" literature or propaganda? (For that matter, is Chiquita Banana an advertising jingle or charming light opera?) But the fuzzy boundary doesn't alter the fact that in literature (as in mathematics) the pure and the applied are different in intent, in method, and in criterion of success.

Perhaps the closest analogy is between mathematics and painting. The origin of painting is physical reality, and so is the origin of mathematics—but the painter is not a camera and the mathematician is not an engineer. The painter of "Uncle Sam Wants You" got his reward from patriotism, from increased enlistments, from winning the war—which is probably different from the reward Rembrandt got from a finished work. How close to reality painting (and mathematics) should be is a delicate matter of judgment. Asking a painter to "tell a concrete story" is like asking a mathematician to "solve a real problem." Modern painting and modern mathematics are far out—too far in the judgment of some. Perhaps the ideal is to have a spice of reality always present, but not to crowd it the way descriptive geometry, say, does in mathematics, and medical illustration, say, does in painting.

Talk to a painter (I did) and talk to a mathematician, and you'll be amazed at how similarly they react. Almost every aspect of the life and of the art of a mathematician has its counterpart in painting, and vice versa. Every time a mathematician hears "I could never make my checkbook balance" a painter hears "I could never draw a straight line"—and the comments are equally relevant and equally interesting. The invention of perspective gave the painter a useful technique, as did the invention of 0 to the mathematician. Old art is as good as new; old mathematics is as good as new. Tastes change, to be sure, in both subjects, but a twentieth century painter has sympathy for cave paintings and a twentieth century mathematician for the fraction juggling of the Babylonians. A painting must be painted and then looked at; a theorem must be printed and then read. The painter who thinks good pictures, and the mathematician who dreams beautiful theorems are dilettantes; an unseen work of art is incomplete. In painting and in mathematics there are some objective standards of good—the painter speaks of structure, line, shape, and texture, where the mathematician speaks of truth, validity, novelty, generality—but they are relatively the easiest to satisfy. Both painters and mathematicians debate among themselves whether these objective standards should even be told to the young—the beginner may misunderstand and overemphasize them and at the same time lose sight of the more important subjective standards of goodness. Painting and mathematics have a history, a tradition, a growth. Students, in both subjects, tend to flock to the newest but, except the very best, miss the point; they lack the vitality of what they imitate, because, among other reasons, they lack the experience based on the traditions of the subject.

I've been talking *about* mathematics, but not *in* it, and, consequently, what I've been saying is not capable of proof in the mathematical sense of the word. I hope just the same, that I've shown you that there is a subject called mathematics (mathology?), and that that subject is a creative art. It is a creative art because mathematicians live, act, and think like artists; and it is a creative art because mathematicians regard it so. I feel strongly about that, and I am grateful for this opportunity to tell you about it. Thank you for listening.

Questions on Halmos

1. Is Halmos saying that mathology is "pure" mathematics, and that mathophysics is "applied"? If not, what is he saying?
2. What is Halmos' definition of mathematics?
3. What does a mathematician mean when he says something is impossible?
4. What is art and why should mathematics be considered a creative art?
5. Discuss theory-builders and problem-solvers in relationship to mathology and mathophysics.
6. What is the measure of goodness in mathematics?

The Meaning of Mathematics

Morris Kline

Mathematics is undoubtedly one of man's greatest intellectual achievements. In addition to the knowledge which the subject itself offers, its language, processes and theories give science its organization and power. Mathematical calculations dictate engineering design. The method of mathematics has inspired social and economic thought, while mathematical thinking has fashioned styles in painting, architecture and music. Even national survival depends today upon progress in mathematics. Finally, mathematics has been a major force in molding our views of the universe and of man's place and purpose in it.

The paradox of how such an abstract body of thought can give man an ever-widening and deepening grip on the physical world and work its influences on almost all phases of our culture tantalizes the nonmathematician. We propose therefore to examine the nature of mathematics and to see why the subject possesses such astonishing effectiveness.

The distinguishing feature of mathematics is its method of reasoning. By measuring the angles of a dozen or so triangles of various shapes and sizes a person would find that the sum in any triangle is 180 degrees. He could then conclude by inductive reasoning that the sum of the angles in every triangle is 180 degrees. One can also reason by analogy. The circle plays about the same role among curves that the sphere does among surfaces. Since the circle bounds more area than any other curve with the same perimeter, a person might conclude that the sphere bounds more volume than any other surface with the same area.

From *Saturday Evening Post*, September 3, 1960. Reprinted by permission of the *Saturday Evening Post* and the author, © 1960, by the Curtis Publishing Company.

Reasoning by induction and by analogy calls for recourse to observation and even experiment to obtain the facts on which to base each argument. But the senses are limited and inaccurate. Moreover, even if the facts gathered for the purpose of induction and analogy are sound, these methods do not yield unquestionable conclusions. For example, though cows eat grass and pigs are similar to cows, it does not follow that pigs eat grass.

To avoid these sources of error, the mathematician utilizes another method of reasoning. He may have the fact that $x - 3 = 7$ and wish to find the value of x. He notes that if he adds 3 to both sides of this equation he will obtain $x = 10$. May he perform this step? He knows that equals added to equals give equals. He knows also that by adding 3 to both sides of the original equation he is adding equals to equals. Hence he concludes that the step is justified. The reasoning here is deductive. As in the present case, so in all deductive reasoning the conclusion is a logically inescapable consequence of the known facts. Hence it is as indubitable as these facts.

Since deduction yields conclusions as certain as the initial facts, the application of this process to known truths produces new ones. The latter may then be used as the premises of new deductive arguments. Every conclusion so obtained may not be significant, but the end result of ten or twenty such arguments could be. If so, it is labeled a theorem. The series of deductive arguments which lead to the theorem is the proof.

Though mathematical proof is necessarily deductive, the creative process practically never is. To foresee what to prove or what chain of deductive arguments will establish a possible result, the mathematician uses observation, measurement, intuition, imagination, induction, or even sheer trial and error. The process of discovery in mathematics is not confined to one pattern or method. Indeed, it is in part as inexplicable as the creative act in any art or science.

The requirement that mathematical reasoning be deductive was laid down by the Greeks. The Greek mathematicians were also philosophers and as such were concerned with truths. They saw clearly that only deductive reasoning could supply certainties. By recommending mental exploration of the riches contained in some available truths, a most reasonable people carved out a new intellectual world and made reason a vital factor in western culture.

The plan to obtain truths by deduction presupposes some initial truths. These the Greeks found in the domains of number and geometrical figures. It seemed axiomatic that equals added to equals should yield equals, that the whole is greater than its parts and that two points determine a straight line. Hence mathematics was built on the axioms of number and geometry. Mathematicians as mathematicians do not reason about forces, weights, sound, light, chemical mixtures or the goal of life.

There were other reasons for the decision to concentrate on number and geometrical figures. The triangle formed by a piece of land and the triangle formed by the earth, sun and moon at any instant are both subsumed under the abstract geometrical concept of triangle. Study of the properties of this concept would yield knowledge about these two physical triangles and about hundreds of others in one swoop. What the Greeks saw, in other words, was that number, size and shape are fundamental properties. In fact, the Greeks believed that the universe was mathematically designed, and so the phenomena of nature could be understood only in terms of number and geometry.

The third feature of mathematics is its highly symbolic language. There is, however, nothing deep or complicated about this language, for it is only a shorthand and, in fact, an easier one to learn that that employed by stenographers. Such symbols as $+$ for addition, x for an unknown quantity, and x^2 for x times x, are, of course, well known. Letters are used for several purposes, and the context usually tells us what is intended. Suppose, for example, we take a famous mathematical statement which describes the result of some experiments made by Galileo about 350 years ago: The number of feet which an object falls in any given number of seconds is 16 times the square of the number of seconds it has been falling. Symbolically this statement is written as $d = 16t^2$ wherein t stands for any number of seconds and d the corresponding distance fallen in these t seconds. Thus, if an object falls for 5 seconds, simple arithmetic shows that it has fallen 400 feet.

Why is symbolism used so extensively? Brevity, precision and comprehensibility are the three major reasons. The brevity is apparent. Precision is aided because many important words of ordinary discourse are ambiguous. The word "equal," for example, can refer to equality in size, shape, political right, intellectual abilities, or other qualities. Hence the assertion that all men are born equal is vague. As used in an expression such as $d = 16t^2$, the equals sign stands for numerical equality. The comprehensibility gained through symbolism derives largely from the fact that the mind easily carries and works with symbolic expressions, but has considerable difficulty even in carrying the equivalent verbal statement.

Our discussion of the method of proof, subject matter and language of mathematics gives some indication of its nature. It is but a step from this point to see some of the sources of the power of mathematics. Number and geometrical figures, and the relationships built on these abstractions, such as formulas, embody the essence of hundreds of physical situations. Any knowledge acquired about these abstractions is many times more potent than that acquired about any particular situation, just as any fact applicable to all men is more powerful than a fact about John Jones. A second source of strength derives from the reliability of deductive proof. Hence the conclusions derived by the Greeks are still acceptable as logical consequences of the axioms and will be a thousand years from now.

The Meaning of Mathematics

But the power of mathematics rests on still another ground. The mathematician is the professional reasoner who devotes his life to learning what has been accomplished in his subject and to extending the results by new reasoning. Moreover, all of the results obtained by one generation are passed on to the next, and this one carries on from where the preceding generation left off. Each generation adds a story to the structure.

To appreciate the full power of mathematics we must examine its role in science. In the seventeenth century, when modern science was founded and received its first great impetus, several major physical laws were obtained by induction and experimentation. Of these we shall be concerned with the second of Newton's three laws of motion and with Newton's law of gravitation. These laws involve a few, by now, common concepts—force, mass, and acceleration. Newton's second law of motion states that any force applied to a mass gives it an acceleration, and the quantitative relation among the amount of force F, the amount of mass m, and the amount of acceleration a is

(1) $\quad F = ma$

The Newtonian law of gravitation states that any two pieces of matter in the universe exert a force of attraction or gravitational force on each other and that the quantitative expression for this force is given by the formula

(2) $\quad F = \dfrac{GmH}{r^2}$

In this equation F is the amount of force exerted; m is the amount of mass in one body; M is the amount of mass in the second body; r is the distance between these bodies; and G is a constant, that is, the same quantity no matter which masses are involved and whatever the distance between them.

These laws concern force, mass and acceleration, which are physical concepts, and the obtainment of relationships among such quantities is the task of the scientist. However, formulas (1) and (2), regarded in and for themselves, are merely algebraic equations relating variables, and it is legitimate to ask the mathematician whether he can draw upon his stock of theorems and processes to deduce new significant equations from (1) and (2). He can. He observes first of all that it is mathematically correct to write formula (2) in the form

(3) $\quad F = m\dfrac{(GM)}{(r^2)}$

He then compares (1) and (3) and observes that the two formulas have the same algebraic form. Moreover, formula (1) applies to any force and so, in particular, applies to the force of gravitation. Since the quantity which multiplies m in (1) is acceleration, the quantity which multiplies m in (3) must also be acceleration. That is, the acceleration which the gravitational force F between M and m imparts to m is

(4) $\quad a = \dfrac{GM}{r^2}$

Next let us apply this formula to a particular situation. Let M denote the mass of the earth. Equation (4) now gives the acceleration which the gravitational force of the earth imparts to any other mass, the acceleration which causes the mass to fall if released from a point above the surface of the earth. It is a fact that the earth acts as if all its mass were concentrated at the center. Then for objects near the surface of the earth the quantity r in (4) is the radius of the earth. The quantity G, as noted above, is a constant under all conditions. Hence, all quantities on the right side of (4) are constant no matter which mass m near the surface of the earth is involved. We may conclude that all bodies fall to earth with the same constant acceleration, a famous result which Galileo discovered experimentally, but which we have deduced from the second law of motion and the law of gravitation.

Now, says the mathematician, we can go a step further. If the quantities a, G, and r in (4) are known, then (4) may be regarded as a simple first-degree equation in the unknown M, and M can readily be calculated. For, correct algebraic steps yield that the mass M of the earth is

(5) $\quad M = \dfrac{ar^2}{G}$

Let us see if we do know the several quantities which appear on the right side of (5). Since the acceleration of all bodies falling to earth is the same, one could take any falling body and measure its acceleration. This quantity had been measured by Galileo and is 32 feet per second each second. The quantity G is a constant under all conditions. It can be and has been measured many times in a laboratory where the conditions are at the experimenter's convenience. Its value is 1.07 divided by 1,000,000,000. The value of the earth's radius r can be determined by a simple application of geometry and was first obtained by the Greek Eratosthenes about 250 B.C. This radius is 4,000 miles or $4,000 \times 5,280$ feet. After the known numerical values are substituted for a, G, and r in (5) one finds that

(6) $\quad M = 131 \times 10^{23}$ pounds.

(The symbol 10^{23} stands for the product $10 \times 10 \times 10 \ldots$ containing 23 factors.) The mass of the earth is a staggering number, but what is more staggering is how easily one finds it.

With essentially such simple tools, Isaac Newton, his contemporaries and his immediate successors calculated the masses of the sun and the several planets, the paths of comets, the motion of the moon and the rise and fall of the tides. In particular, Newton showed that the Keplerian laws of planetary motion, which Kepler had obtained merely by induction from data, were logical consequences of the laws of motion and the law of gravitation. Thus the key laws of the heliocentric theory of planetary motion, which up to that time were unrelated to any basic physical principles, received indisputable support.

The work we have just described belongs to celestial mechanics, a field which has come to the fore again to treat the motion of satellites. It was followed by the construction of equally majestic theories for light, sound, electromagnetic waves (which comprise the radio waves, the existence of which was predicted mathematically): the flow of fluids and gases as applied to the design of ships and airplanes; relativity, atomic structure, molecular structure (now basic in modern chemistry); the biological science of mathematical genetics and the statistical treatment of social and medical problems. In all these domains the union of mathematics and science has been most fertile.

The contributions of a special, functional language and the deductive processes are a small part of the mathematical largess. Science seeks to obtain knowledge of the physical world, but that knowledge would be useless if unorganized. A mass of disconnected results is no more science than a collection of bricks is a house. The major results of scientific work are theories. In each of these, hundreds of results are organized in a deductive structure very much like Euclidean geometry. At the head of the structure are basic physical principles which play the role of axioms. From these axioms the various laws of any one theory are deduced. The large and over-riding fact is that the entire structure of a scientific theory is held together by a series of mathematical deductions. The mortar which binds the bricks, or individual laws, one to another is mathematical deducibility. A scientific theory is, so to speak, a branch of mathematics whose axioms state quantitative relationships among physical concepts, whose structure is a series of mathematical deductions and whose theorems are mathematical affirmations about these concepts.

Mathematics plays still another role in science. The central concept in the most impressive and most successful body of science, mechanics, is the force of gravitation. This force, when exerted by the earth, pulls objects to the earth and, when exerted by the sun, keeps the planets in their paths. What is the mechanism by which the earth and the sun exert their respective attractive forces? Newton had considered this very question and, having failed to answer it, uttered his famous "I frame no hypotheses." The history of this subject is extensive, but the upshot of it is that no explanation of the action of the force of gravitation has ever been given.

What then do we know about the force of gravitation? The answer is formula (2) above. We have a quantitative law which tells us how to calculate this force and from which we can deduce how bodies will move, what paths they will take, and where they will be at a particular instant of time. We have not a shred of insight into the physical nature of the force itself; it can with full justification be regarded as sheer fiction.

In the Newtonian age mathematics mounted the steed of science and took the reins in its own hands. Since the seventeenth century the physical behavior of nature has become less and less clear despite the vast expansion of the sciences, and mathematical laws have become the essence and goal of science. The mathematical conquest of science has by now proceeded so far that in our own century the late Sir James Jeans, the noted astronomer and physicist, claimed that the mathematical description of the universe is the ultimate reality. The pictures and models we use to assist our understanding are a step away. We go beyond the mathematical formula at our own risk.

While mathematical physics was growing to manhood, mathematics began to exert a formative influence on numerous other branches of our culture. Revival of interest in the physical world caused the Renaissance painters to abandon the unrealistic, highly symbolic style of medieval painting and to seek a veridical depiction of nature. To solve the problem of presenting on a flat surface scenes which would create the same visual impression as the three-dimensional world itself, the painters created a mathematical system of perspective painting. The introduction of depth, solidity, mass and consequent realism is the key contribution to Renaissance painting.

The mathematical treatment of matter in motion engendered the now famous philosophical doctrines that every phenomenon in the universe can be reduced to matter and motion, that all matter, including man's will, is bound fast, and that thought is but a mechanical reaction to material sensations impressed on the brain through the sense organs. Inspired by the success of the mathematical method in the physical sciences and enthusiastic about the power of reason exercised through mathematics, leading eighteenth-century thinkers undertook a rational approach to social problems and launched the sciences of government and economics. The spread of this same rational spirit freed man from superstitions and groundless fears and permitted him to breathe in a more tolerant atmosphere.

The proliferating demands of science, which first became urgent during the seventeenth century, stimulated an enormous expansion in mathematics proper. To obtain a deeper appreciation of the nature of modern mathematics we must look into these more recent developments. The mathematics which the Europeans possessed by 1600 consisted of algebra, Euclidean geometry and the beginnings of trigonometry. In the seventeenth century the need to study curves—whether the paths of light through lenses, the paths of cannonballs, the paths of ships at sea, or the paths of the planets—prompted René Descartes and Pierre de Fermat to create an algebraic method of representing curves so that algebra could be used to deduce the properties of curves. This creation is known as co-ordinate or analytic geometry.

The need to calculate varying velocity, force, pressure, and other physical quantities in problems of celestial mechanics as well as in navigation and gunfire was met by the creation of a new concept, the concept of a limit, and a new method called differentiation. This is the substance of the differential calculus. To obtain the sum of an infinite number of small quantities, for example, the sum of the gravitational forces which each bit of earth exerts on some external mass, the integral calculus was created.

The calculus was the beginning of a series of new branches commonly grouped under the name of analysis. Differential equations, infinite series, the differential geometry, the calculus of variations, functions of a complex variable and vector and tensor analysis are but a few of the subdivisions of analysis. The domain of algebra was likewise extended to include such abstractions as complex numbers, vectors, hypernumbers, matrices, abstract sets and the theory of structures of algebra known as abstract algebra. Projective geometry, non-Euclidean geometry, algebraic geometry and topology joined Euclidean geometry.

The major motivation for all these creations was to further the leading physical studies of the eighteenth and nineteenth centuries—the strength of beams in structures, the motion of ships, the flow of the tides, the development of steam as a source of power, the generation and utilization of electrical power, the improvement of optical instruments, ballistics, and dozens of other new or growing scientific interests. But we should not overlook the fact that mathematicians enjoy the creative mathematical activity itself, the intellectual challenge, the satisfaction of accomplishment and the beauty of proofs and results. Given the themes suggested by physical problems, mathematicians develop these far beyond the needs of science, often to find that they have anticipated other needs or have unintentionally supplied the concepts and frameworks for new physical theories.

While the proliferation of mathematics is a phenomenon of our modern culture, an even more startling development has been the realization that mathematics is not an absolute, all-embracing truth, or description of reality, in the sense that man had until recently thought it was. For 2000 years the axioms of number and geometry were accepted as self-evident truths. Since the theorems are logically necessary consequences of the axioms, the theorems, too, were believed to be incontrovertible truths.

The creation of non-Euclidean geometry had the unintended effect of thrusting mathematics off this pedestal. Historically, non-Euclidean geometry was the result of attempts to find a simpler version of the Euclidean parallel-line axiom which postulates, in effect, that through any point in a given plane there is one, and only one, line parallel to a given line. In the course of this research, mathematicians deliberately adopted an axiom on parallel lines which contradicted Euclid's axiom. From this new axiom and the remainder of Euclid's axioms they proceeded to deduce theorems. They expected to arrive at inconsistencies within the new geometry; that is, they expected to find some theorems contradicting others because they had started with an axiom, which so they thought, denied the truth. But these contradictions failed to appear!

The supreme mathematician of the nineteenth century, Karl Friedrich Gauss, was the first to see the handwriting on the wall. He realized that Euclidean geometry could no longer be regarded as the only geometry of physical space and that non-Euclidean geometry might do as well. Further, his efforts to test experimentally which of the geometries, Euclidean or non-Euclidean, fits the physical world better, ended in failure. The situation became even more unsettled when Bernhard Riemann created additional non-Euclidean geometries. The potential applicability of all non-Euclidean geometries was increased when mathematicians recognized that the physical "straight" lines used in the most weighty scientific work are not stretched strings or rulers' edges but paths of light rays. Since these paths are generally not straight, the geometry whose axioms fitted their behavior and the behavior of figures formed by such "lines" could very well be one of the non-Euclidean varieties. The mathematicians were ultimately forced to admit that there was no reason to believe in the exclusive truth of any one of these geometries. When the theory of relativity made use of a non-Euclidean geometry, the point was driven home.

Some mathematicians sought refuge in those portions of mathematics which rest on the number system and maintained that these at least offer truths. However, this thesis is also indefensible, for we now see more clearly that while the arithmetic we ordinarily use fits the common situations involving quantity, there are other arithmetics and their algebras which fit other situations. To mention a trivial example, an alternative arithmetic fitting a real situation is used when we state that four hours after nine o'clock the hour will be one o'clock rather than thirteen o'clock.

A word of comfort here to the non-mathematician who fears that he may have learned the rudiments of arithmetic and geometry in vain—or suspects that my earlier statements concerning the validity of mathematical processes are now contradicted. "Two plus two equals four" is still a valid deduction from the axioms of arithmetic just as the theorems of Euclidean geometry are still valid deductions from Euclid's axioms. However, the arithmetic and geometric conclusions can be applied only where experience tells us that the axioms are applicable. Thus we shall still use the fact that 2 dollars plus 2 dollars are 4 dollars, but not that 2 raindrops added to 2 raindrops are 4 raindrops. Two raindrops plus 2 raindrops make a puddle. Again, if we mix 2 cubic inches of hydrogen and 1 of oxygen, we obtain not 3 but 2 cubic inches of water. Philosophically this suggests that "truth" in mathematics, as in all human processes, is a many-faceted thing.

Recognition of the shattering fact that mathematics, which had always been regarded as the anchor of truth and as conclusive evidence that man can attain truths, rests on pragmatic grounds was a direct result of the nineteenth-century questioning of man's assumptions concerning the physical world. Mathematicians had believed that the assumptions or axioms—and therefore the logical consequences—were truths. It was now realized that such axioms are man-made inferences based on limited sense data and are only approximations of what happens in the physical world. In fact, the word "axiom" should now be taken to mean assumption, rather than self-evident truth. We continue to use the axioms and conclusions, even though they are not truths, because they do offer some highly useful knowledge about the physical world—the best knowledge, in fact, that man possesses.

Oddly though, when truth, the most prized possession of mathematics, was taken from it, the subject emerged richer for the loss. Axioms palpably untrue had led to geometries which proved useful. This experience justified the exploration of any system of axioms, however unpromising for application it might seem at the outset. Mathematics, which had been fettered to the physical world, passed from serfdom to freedom.

There is no doubt of the positive value of the new freedom. From the unrestricted play of mathematical imagination have come and will continue to come systems of thought which may prove to be far more valuable in representing and mastering the physical world than could have come from concentration on the two original systems of number and Euclidean geometry. So it was, when Einstein needed to know the structure of a particular four-dimensional, non-Euclidean geometry, that he found the information at hand.

Examination of newer mathematics reveals another gradual change in its nature. The early concepts of mathematics, the whole number, the fractions and several geometrical figures, were clearly suggested by immediate experiences. Mathematicians later found themselves developing and applying such abstract extensions of the idea of number as irrational, negative and complex numbers. Because they did not at first understand these new types of numbers or recognize their usefulness, even the greatest ones resisted their introduction. Having worked for centuries with numbers and geometrical figures which were suggested directly by physical objects, mathematicians had implicitly and unconsciously concluded that their concepts must be "real." What was involved in the acceptance of new types of numbers was a sharp break from concepts grounded directly in experience. Mathematicians have since learned the deeper meaning of the statement that mathematics is an activity of the human mind and now grant that any concept which is clear and fertile should be explored whether or not it has an apparent physical basis.

Paradoxically, to obtain insights into the physical world, we must plunge deeper into human minds, consider abstractions that are remote from reality, and explore the implications of axioms that not only transcend but even appear to deny our sense impressions. Though recourse to higher and higher abstractions seems regrettable, the case for it was superbly stated by the distinguished philosopher, Alfred North Whitehead: "Nothing is more impressive than the fact that as mathematics withdrew increasingly into the upper regions of ever greater extremes of abstract thought, it returned to earth with a corresponding growth of importance for the analysis of concrete fact . . . . The paradox is now fully established that the utmost abstractions are the true weapons with which to control our thought of concrete fact."

The pleasures and pains of mathematical activity have been recommended largely on the ground that they help us to achieve knowledge of the physical world. Why do we seek this knowledge? The ultimate goal of scientific activity is man himself. He wants to know the meaning of his own life and seeks the answer by attempting to understand the world in which he finds himself.

Mathematics mediates between man and nature, between man's inner and outer worlds. What mathematical concepts and methods have achieved in rationalizing nature have yielded our clearest and most weighty scientific doctrines. When solely for mathematical reasons Copernicus and Kepler adopted a new mathematical scheme by which to organize the observations of the heavens and place the sun rather than the earth in the center, they caused man to recognize that he was an insignificant creature whirling through vast spaces rather than the central figure in the drama of nature. When Newtonian mechanics revealed a universe firmly controlled by definite mathematical laws and functioning both in the past and in the future to no end other than the fulfillment of mathematical laws, man had to cope with the implication that he was without will or purpose. If more recent creations such as quantum theory have cast in doubt the bleak, mechanical, deterministic implications of earlier theories and have given man some hope of reinstatement in an important role, it is still true that his outlook is confined and directed by mathematical chains of thought.

Science provides the understanding of the universe in which we live. Mathematics provides the dies by which science is molded. Our world is to a large extent what mathematics says it is. The body of man-made abstractions, wherein, as Bertrand

Russell put it, we never know what we are talking about nor whether what we are saying is true, this practical tool, model of all intellectual enterprises, and essence of our knowledge of nature leads through science to man himself.

Questions on Kline

1. How does Kline distinguish between inductive and deductive reasoning?
2. Is mathematical proof a deductive or inductive process?
3. How does Kline define theorem and proof?
4. What did the Greeks see as fundamental properties of things around them?
5. What features of mathematics does Kline consider important?
6. Why is mathematics so symbolic?
7. What motivates mathematicians, and where do they find their problems?
8. What does Kline say about mathematics being a description of reality? Of absolute truth?
9. Comment on "Science provides the understanding of the universe in which we live. Mathematics provides the dies by which science is molded."
10. Discuss the role of mathematics as the "mediator between man's inner and outer world."

Mathematics as an Element in the History of Thought

Alfred North Whitehead

The science of pure mathematics, in its modern developments, may claim to be the most original creation of the human spirit. Another claimant for this position is music. But we will put aside all rivals, and consider the ground on which such a claim can be made for mathematics. The originality of mathematics consists in the fact that in mathematical science connections between things are exhibited which, apart from the agency of human reason, are extremely unobvious. Thus the ideas, now in the minds of contemporary mathematicians, lie very remote from any notions which can be immediately derived by perception through the senses: unless indeed it be perception stimulated and guided by antecedent mathematical knowledge. This is the thesis which I proceed to exemplify.

Suppose we project our imagination backwards through many thousands of years, and endeavour to realise the simple-mindedness of the greatest intellects in those early societies. Abstract ideas which to us are immediately obvious must have been, for them, matters only of the most dim apprehension. For example take the question of number. We think of the number "five" as applying to appropriate groups of any entities whatsoever—to five fishes, five children, five apples, five days. Thus in considering the relations of the number "five" to the number "three," we are thinking of two groups of things, one with five members and the other with three members. But we are entirely abstracting from any consideration of any particular entities, or even of any particular sorts of entities, which go to make up the membership of either the two groups. We are merely thinking of those relationships between

---

Reprinted with permission of the Macmillan Company and Cambridge University Press from *Science and the Modern World* by A. N. Whitehead. Copyright 1925 by the Macmillan Company, renewed 1953 by Evelyn Whitehead.

those two groups which are entirely independent of the individual essences of any of the members of either group. This is a very remarkable feat of abstraction; and it must have taken ages for the human race to rise to it. During a long period, groups of fishes will have been compared to each other. But the first man who noticed the analogy between a group of seven fishes and a group of seven days made a notable advance in the history of thought. He was the first man who entertained a concept belonging to the science of pure mathematics. At that moment it must have been impossible for him to divine the complexity and subtlety of these abstract mathematical ideas which were waiting for discovery. Nor could he have guessed that these notions would exert a wide-spread fascination in each succeeding generation. There is an erroneous literary tradition which represents the love of mathematics as a monomania confined to a few eccentrics in each generation. But be this as it may, it would have been impossible to anticipate the pleasure derivable from a type of abstract thinking which had no counterpart in the then-existing society. Thirdly, the tremendous future effect of mathematical knowledge on the lives of men, on their daily avocations, on their habitual thoughts, on organization of society, must have been even more completely shrouded from the foresight of those early thinkers. Even now there is a very wavering grasp of the true position of mathematics as an element in the history of thought. I will not go so far as to say that to construct a history of thought without profound study of the mathematical ideas of successive epochs is like omitting Hamlet from the play which is named after him. That would be claiming too much. But it is certainly analogous to cutting out the part of Ophelia. This simile is singularly exact. For Ophelia is quite essential to the play, she is very charming—and a little mad. Let us grant that the pursuit of mathematics is a divine madness of the human spirit, a refuge from the goading urgency of contingent happenings.

When we think of mathematics, we have in our mind a science devoted to the exploration of number, quantity, geometry, and in modern times also including investigation into yet more abstract concepts of order, and into analogous types of purely logical relations. The point of mathematics is that in it we have always got rid of the particular instance, and even of any particular sorts of entities. So that for example, no mathematical truths apply merely to fish, or merely to stones, or merely to colours. So long as you are dealing with pure mathematics, you are in the realm of complete and absolute abstraction. All you assert is, that reason insists on the admission that, if any entities whatever have any relations which satisfy such-and-such purely abstract conditions, then they must have other relations which satisfy other purely abstract conditions.

Mathematics is thought moving in the sphere of complete abstraction from any particular instance of what it is talking about. So far is this view of mathematics from being obvious, that we can easily assure ourselves that it is not, even now, generally understood. For example, it is habitually thought that the certainty of mathematics is a reason for the certainty of our geometrical knowledge of the space of the physical universe. This is a delusion which has vitiated much philosophy in the past, and some philosophy in the present. The question of geometry is a test case of some urgency. There are certain alternative sets of purely abstract conditions possible for the relationship of groups of unspecified entities, which I will call geometrical conditions. I give them this name because of their general analogy to those conditions, which we believe to hold respecting the particular geometrical relations of things observed by us in our direct perception of nature. So far as our observations are concerned, we are not quite accurate enough to be certain of the exact conditions regulating the things we come across in nature. But we can by a slight stretch of hypothesis identify these observed conditions with some one set of the purely abstract geometrical conditions. In doing so, we make a particular determination of the group of unspecified entities which are the relata in the abstract science. In the pure mathematics of geometrical relationships, we say that, if any group of entities enjoy any relationships among its members satisfying this set of abstract geometrical conditions, then such-and-such additional abstract conditions must also hold for such relationships. But when we come to physical space, we say that some definitely observed group of physical entities enjoys some definitely observed relationships among its members which do satisfy this above-mentioned set of abstract geometrical conditions. We thence conclude that the additional relationships which we conclude to hold in any such case, must therefore hold in this particular case.

The certainty of mathematics depends upon its complete abstract generality. But we can have no certainty that we are right in believing that the observed entities in the concrete universe form a particular instance of what falls under our general reasoning. To take another example from arithmetic, it is a general abstract truth of pure mathematics that any group of forty entities can be subdivided into two groups of twenty entities. We are therefore justified in concluding that a particular group of apples which we believe to contain forty members can be subdivided into two groups of apples of which each contains twenty members. But there always remains the possibility that we have miscounted the big group; so that, when we come in practice to subdivide it, we shall find that one of the two heaps has an apple too few or an apple too many.

Accordingly, in criticising an argument based upon the application of mathematics to particular matters of fact there are always three processes to be kept perfectly distinct in our minds. We must first scan the purely mathematical reasoning to make sure that there are no mere slips in it—no casual illogicalities due to mental failure. Any mathematician knows from bitter experience that in first elaborating a train of reasoning, it is very easy to commit a slight error which yet makes all the difference. But when a piece of mathematics has been revised, and has been before the expert world for some time, the chance of a casual error is almost negligible. The next process is to make quite certain of all the abstract conditions which have been presupposed to hold. This is the determination of the abstract premises from which the mathematical reasoning proceeds. This is a matter of considerable difficulty. In the past quite remarkable oversights have been made, and have been accepted by generations of the greatest mathematicians. The chief danger is that of oversight, namely, tacitly to introduce some condition, which it is natural for us to presuppose, but which in fact need not always be holding. There is another opposite oversight in this connection which does not lead to error, but only to lack of simplification. It is very easy to think that more postulated conditions are required than is in fact the case. In other words, we may think that some abstract postulate is necessary which is in fact capable of being proved from the other postulates that we have already on hand. The only effects of this excess of abstract postulates are to diminish our aesthetic pleasure in the mathematical reasoning, and to give us more trouble when we come to the third process of criticism.

This third process of criticism is that of verifying that our abstract postulates hold for the particular case in question. It is in respect to this process of verification for the particular case that all the trouble arises. In some simple instances, such as the counting of forty apples, we can with a little care arrive at practical certainty. But in general, with more complex instances, complete certainty is unattainable. Volumes, libraries of volumes, have been written on the subject. It is the battle ground of rival philosophers. There are two distinct questions involved. There are particular definite things observed, and we have to make sure that the relations between these things really do obey certain definite exact abstract conditions. There is great room for error here. The exact observational methods of science are all contrivances for limiting these erroneous conclusions as to direct matters of fact. But another question arises. The things directly observed are, almost always, only samples. We want to conclude that the abstract conditions, which hold for the samples, also hold for all other entities which, for some reason or other, appear to us to be of the same sort. This process of reasoning from the sample to the whole species is Induction. The theory of Induction is the despair of philosophy—and yet all our activities are based upon it. Anyhow, in criticising a mathematical conclusion as to a particular matter of fact, the real difficulties consist in finding out the abstract assumptions involved, and in estimating the evidence for their applicability to the particular case in hand.

Questions on Whitehead

1. What does Whitehead think mathematics is? Compare his view with those of Halmos and Kline.
2. What is meant by mathematical certainty? Compare this with impossibility as Halmos discusses it.
3. What different kinds of truth does Whitehead discuss? Briefly discuss them in your own words.
4. Discuss the role that "self-evident" truths play in mathematics. Draw freely upon each of the three articles.
5. What does Whitehead mean by a postulate? Of what importance are they?
6. Summarize Whitehead's ideas concerning inductive and deductive reasoning and the role of each in thinking.

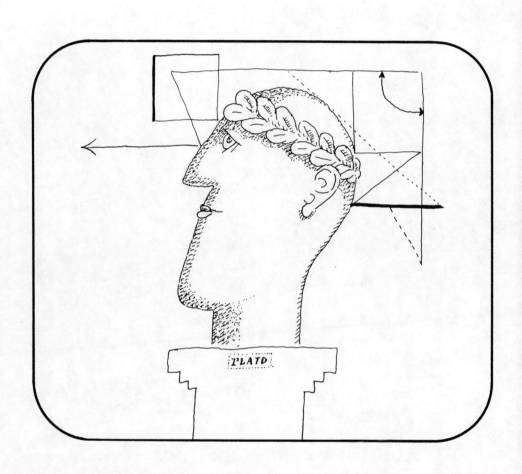

Part Two

Some Historical Facts and Philosophical Beginnings

Introduction

The first selection in this section gives you a listing of some of the important recorded events in mathematics, along with the usual events that are called historical.

The first of the two excerpts from Plato is a continuation of the selection that formed the introduction to this book. It is not clear whether or not Plato should be considered a mathematician. Internal evidence contained in the dialogues shows that:

1. Plato knew quite a lot of the mathematics of his day.
2. He knew what problems were being raised concerning some of that mathematics.
3. He liked mathematics and encouraged its study.
4. He felt there was a kind of truth to be obtained by thinking mathematically that could be gotten in no other way.

So-called Platonic realism has persisted in many guises through the centuries since Plato. Despite many attempts to abandon Plato in mathematics, we continually find ourselves spiralling back and rearticulating something Plato said over 2300 years ago. Book VII of the *Republic* provides a lens through which all of Plato's other thoughts are focused. In this book the Theory of Forms of reality is formally enunciated and expositorily presented. One should also note the major role played by mathematics in the education of the rulers.

As aids to complete understanding, you might keep the following beliefs of Plato in mind. You may find in your reading that you disagree with them. If you do, discuss with someone what ideas you would put in their place.

1. Plato assumes man can tell the difference between appearance and reality or can make that distinction clear.
2. Plato seems to admit only two kinds of absolutely real Forms, mathematical and moral.

3. For Plato, the world of Forms is timeless, independent of mind. It consists of definite objects but is distinct from the world of sense perception. The world of Forms is apprehended not by the senses but by the soul or reason.

4. According to Plato, "pure" mathematics describes empirical objects and their relations in so far as they approximate or participate in the mathematical Forms and relations between Forms.

A Chronology of Mathematical and Nonmathematical Events*

Carl B. Boyer

		−5,000,000,000,000	Origin of the sun
		−5,000,000,000	Origin of the earth
		−600,000,000	Beginning of Paleozoic Age
		−225,000,000	Beginning of Mesozoic Age
		−60,000,000	Beginning of Cenozoic Age
		−2,000,000	Origin of man
−50,000	Evidence of counting	−50,000	Neanderthal Man
−25,000	Primitive geometrical designs	−25,000	Paleolithic art; Cro-Magnon Man
		−10,000	Mesolithic agriculture
		−5000	Neolithic civilizations
−4241	Hypothetical origin of Egyptian calendar		
		−4000	Use of metals
		−3500	Use of potter's wheel; writing

* Dates before −776 are approximations only.

Reprinted with permission from *A History of Mathematics* by Carl B. Boyer, copyright 1968 by John Wiley and Sons, Inc.

A Chronology of Mathematical and Nonmathematical Events

−3000	Hieroglyphic numerals in Egypt	−3000	Use of wheeled vehicles
−2773	Probable introduction of Egyptian calendar	−2800	Great Pyramid
−2400	Positional notation in Mesopotamia	−2400	Sumerian-Akkadian Empire
−1850	Moscow (Golenishev) papyrus; cipherization		
		−1800	Code of Hammurabi
		−1700	Hyksos domination of Egypt; Stonehenge in England
		−1600	Kassite rule in Mesopotamia; New Kingdom in Egypt
		−1400	Catastrophe in Crete
		−1350	Phoenician alphabet; use of iron; sundial; water clocks
		−1200	Trojan War; Exodus from Egypt
−1100?	Chou-pei		
		−776	First Olympiad
		−753	Traditional founding of Rome
		−743	Era of Nabonassar
		−740	Works of Homer and Hesoid (approx.)
		−586	Babylonian Captivity
−585	Thales of Miletus; deductive geometry (?)		
−540	Pythagorean arithmetic and geometry (approx.) Rod numerals in China (approx.) Indian *Sulvasūtras* (approx.)		
		−538	Persians took Babylon
		−480	Battle of Thermopylae
		−477	Formation of Delian League
		−461	Beginning of Age of Pericles
−450	Spherical earth of Parmenides (approx.)		
−430	Death of Zeno; works of Democritus Astronomy of Philolaus (approx.) *Elements* of Hippocrates of Chios (approx.)	−430	Hippocrates of Cos (approx.) Atomic doctrine (approx.)
		−429	Death of Pericles; plague at Athens
−428	Birth of Archytas; death of Anaxagoras		
−427	Birth of Plato		
−420	Trisectrix of Hippias (approx.) Incommensurables (approx.)		
		−404	End of Peloponnesian War
		−399	Death of Socrates; *Anabasis* of Xenophon

−369	Death of Theaetetus		
−360	Eudoxus on proportion and exhaustion (approx.)		
−350	Menaechmus on conic sections (approx.) Dinostratus on quadratrix (approx.)		
		−347	Death of Plato
−335	Eudemus: *History of Geometry* (approx.)		
		−332	Alexandria founded
−330	Autolycus: *On the Moving Sphere* (approx.)		
		−323	Death of Alexander
		−322	Deaths of Aristotle and Demosthenes
−320	Aristaeus: *Conics* (approx.)		
		−311	Beginning of Seleucid Era in Mesopotamia
		−306	Ptolemy I (Soter) of Egypt
−300	Euclid's *Elements* (approx.)		
		−283	Pharos at Alexandria
		−264	First Punic War opened
−260	Aristarchus' heliocentric astronomy (approx.)		
		−232	Death of Asoka, the "Buddhist Constantine"
−230	Sieve of Eratosthenes (approx.)		
−225	*Conics* of Apollonius (approx.)		
−212	Death of Archimedes		
		−210	Great Chinese Wall begun
−180	Cissoid of Diocles (approx.) Conchoid of Nicomedes (approx.) Hypsicles and 360° circle (approx.)		
		−166	Revolt of Judas Maccabaeus
−150	Spires of Perseus (approx.)		
		−146	Destruction of Carthage and Corinth
−140	Trigonometry of Hipparchus (approx.)		
		−121	Gaius Gracchus killed
		−75	Cicero restored tomb of Archimedes
−60	Geminus on parallel postulate (approx.)	−60	Lucretius: *De rerum natura*
		−44	Death of Julius Caesar
+75	Works of Heron of Alexandria (approx.)		
		+79	Death of Pliny the Elder at Vesuvius
100	Nicomachus: *Arithmetica* (approx.) Menelaus: *Spherics* (approx.)		
		116	Trajan extends Roman Empire
		122	Hadrian's Wall in Britain begun
125	Theon of Smyrna and Platonic mathematics		

A Chronology of Mathematical and Nonmathematical Events

150	Ptolemy: *The Almagest* (approx.)		
		180	Death of Marcus Aurelius
250	Diophantus: *Arithmetica* (approx.?)		
		286	Division of Empire by Diocletian
320	Pappus: *Mathematical Collections* (approx.)		
		324	Founding of Constantinople
		378	Battle of Adrianople
390	Theon of Alexandria (fl.)		
415	Death of Hypatia		
		455	Vandals sack Rome
470	Tsu Ch'ung-chi's value of π (approx.)	476	Traditional "fall" of Rome
476	Birth of Aryabhata		
485	Death of Proclus		
		496	Clovis adopted Christianity
520	Anthemius of Tralles and Isidore of Miletus		
524	Death of Boethius		
		526	Death of Theodoric
529	Closing of the schools at Athens	529	Founding of the monastery at Monte Cassino
		532	Building of Hagia Sophia by Justinian
560	Eutocius' commentaries on Archimedes (approx.)		
		590	Gregory the Great elected pope
		622	Hejira of Mohammed
628	Brahma-sphuta-siddhânta		
		641	Library at Alexandria burned
662	Bishop Sebokht mentioned Hindu numerals		
		732	Battle of Tours
735	Death of the Venerable Bede		
775	Hindu works translated into Arabic		
		814	Death of Charlemagne
830	Al–Khowarizmi: *Algebra* (approx.)		
901	Death of Thabit ibn-Qurra		
		910	Benedictine abbey at Cluny
		987	Accession of Hugh Capet
998	Death of abu'l-Wefa		
		999	Gerbert became Pope Sylvester II
		1028	School of Chartres
1037	Death of Avicenna		
1039	Death of Alhazen		
1048	Death of al-Biruni		
		1066	Battle of Hastings
		1096	First Crusade
		1100	Henry I of England crowned
1114	Birth of Bhaskara		
1123	Death of Omar Khayyam		

1142	Adelard of Bath translated Euclid		
		1170	Murder of Thomas à Becket
1202	Fibonacci: *Liber abaci*		
		1204	Crusaders sack Constantinople
			Death of Maimonides
		1215	Magna Carta
1260	Campanus trisection (approx.)		
	Jordanus Nemorarius: *Arithmetica* (approx.)		
		1265	"First" parliament in England
1270	Wm. of Moerbeke translated Archimedes (approx.)		
		1271	Travels of Marco Polo; mechanical clocks (approx.)
1274	Death of Nasir Eddin		
		1286	Invention of eyeglasses (approx.)
1303	Chu Shi-kié and the Pascal triangle		
1328	Bradwardine: *Liber de proportionibus*		
1336	Death of Richard of Wallingford		
		1348	The Black Death
1360	Oresme's latitude of forms (approx.)		
		1364	Death of Petrarch
		1431	Joan of Arc burned
1436	Death of al-Kashi		
		1440	Invention of printing
		1453	Fall of Constantinople
1464	Death of Nicholas of Cusa		
1472	Peurbach: *New Theory of the Planets*		
		1473	Sistine Chapel
1476	Death of Regiomontanus		
1482	First printed Euclid		
		1483	Murder of the princes in the Tower
1484	Chuquet: *Triparty*		
		1485	Henry VII, first Tudor
1489	Use of + and − by Widmann		
1492	Use of decimal point by Pellos	1492	Discovery of America by Columbus
1494	Pacioli: *Summa*		
		1498	Execution of Savonarola
		1517	Protestant Reformation
		1520	Field of the Cloth of Gold
1525	Rudolff: *Coss*		
1526	Death of Scipione dal Ferro		
1527	Apian published the Pascal triangle		
		1534	Act of Supremacy
1543	Tartaglia published Moerbeke's *Archimedes*	1543	Vesalius: *De fabrica*
	Copernicus: *De revolutionibus*		Ramus: *Reproof of Aristotle*
1544	Stifel: *Arithmetica integra*		

A Chronology of Mathematical and Nonmathematical Events 55

1545	Cardan: *Ars magna*		
		1553	Servetus burned at Geneva
1557	Recorde: *Whetstone of Witte*		
		1558	Accession of Elizabeth I
1564	Birth of Galileo	1564	Birth of Shakespeare; deaths of Vesalius and Michelangelo
1572	Bombelli: *Algebra*	1572	Massacre of St. Bartholomew
1579	Viète: *Canon mathematicus*		
		1584	Assassination of William of Orange
1585	Stevin: *La disme*		
	Harriot's report on "Virginia"		
		1588	Drake defeated the Spanish Armada
1595	Pitiscus: *Trigonometria*		
		1598	Edict of Nantes
1603	Death of Viète	1603	Deaths of Wm. Gilbert and Elizabeth I
1609	Kepler: *Astronomia nova*	1609	Galileo's telescope
1614	Napier's logarithms		
		1616	Deaths of Shakespeare and Cervantes
1620	Bürgi's logarithms	1620	Landing of the Pilgrims
		1626	Deaths of Francis Bacon and Willebrord Snell
		1628	Harvey: *De motu cordis et sanguinis*
1629	Fermat's method of maxima and minima		
1631	Harriot: *Artis analyticae praxis*		
	Oughtred: *Clavis mathematicae*		
1635	Cavalieri: *Geometria indivisibilibus*		
		1636	Harvard College founded
1637	Descartes: *Discours de la méthode*		
1639	Desargues: *Bruillon projet*		
1640	*Essay pour les coniques* of Pascal		
1642	Birth of Newton; death of Galileo		
		1643	Accession of Louis XIV
		1644	Torricelli's barometer
1647	Deaths of Cavalieri and Torricelli		
		1649	Charles I beheaded
		1651	Hobbes: *Leviathan*
			Von Guericke's air pump
1655	Wallis: *Arithmetica infinitorum*		
1657	Neil rectified his parabola		
1658	Huygen's cycloidal pendulum clock		
		1660	The Restoration
		1662	Royal Society founded
		1666	Académie des Sciences founded
1667	Gregory: *Geometriae pars universalis*		
1668	Mercator: *Logarithmotechnia*		
1670	Barrow: *Lectiones geometriae*		

1672	Assassination of De Witt		
1678	Ceva's theorem		
		1679	Writ of Habeas Corpus
		1682	*Acta eruditorum* founded
		1683	Siege of Vienna
1684	Leibniz' first paper on the calculus		
		1685	Revocation of the Edict of Nantes
1687	Newton: *Principia*		
		1689	The Glorious Revolution
1690	Rolle: *Traité d'algèbre*		
1696	Brachistochrone (the Bernoullis) L'Hospital's rule		
		1699	Death of Racine
		1702	Opening of Queen Anne's War
1706	Use of π by William Jones		
		1711	Birth of Hume
1715	Taylor: *Methodus incrementorum*		
1718	De Moivre: *Doctrine of Chances*	1718	Fahrenheit's thermometer
1722	Cotes: *Harmonia mensurarum*		
1730	Stirling's formula	1730	Réaumur's thermometer
1731	Clairaut on skew curves		
1733	Saccheri: *Euclid Vindicated*		
1734	Berkeley: *The Analyst*		
		1737	Linnaeus: *Systema naturae*
		1738	Daniel Bernoulli: *Hydrodynamica*
		1740	Accession of Frederick the Great
1742	Maclaurin: *Treatise of Fluxions*	1742	Centigrade thermometer
1743	D'Alembert: *Traité de dynamique*		
1748	Euler: *Introductio*; Agnesi: *Istituzioni*		
		1749	Volume I of Buffon's *Histoire naturelle*
1750	Cramer's rule; Fagnano's ellipse		
		1751	Volume I of Diderot's *Encyclopédie*
		1752	Franklin's kite
1759	*Die freye Perspektive* of Lambert		
		1767	Watt's improved steam engine
1770	Hyperbolic trigonometry		
		1774	Discovery of Oxygen (Priestley, Scheele, Lavoisier)
		1776	American Declaration of Independence
1777	Buffon's needle problem		
1779	Bézout on elimination		
		1781	Discovery of Uranus by Herschel
		1783	Composition of water (Cavendish, Lavoisier)
1788	Lagrange: *Mécanique analytique*		
		1789	French Revolution

A Chronology of Mathematical and Nonmathematical Events

1794	Legendre: *Eléments de géométrie*	1794	Lavoisier guillotined
1795	Monge: *Feuilles d'analyse*	1795	Ecole Polytechnique: Ecole Normale
1796	Laplace: *Système du monde*	1796	Vaccination (Jenner)
1797	Lagrange: *Fonctions analytiques*		
	Mascheroni: *Geometria del compasso*		
	Wessel: *Essay on ... direction*		
	Carnot: *Métaphysique du calcul*		
		1799	Metric system
		1800	Volta's battery
1801	Gauss: *Disquisitiones arithmeticae*	1801	Ceres discovered
1803	Carnot: *Géométrie de position*	1803	Dalton's atomic theory
		1804	Napoleon crowned emperor
1810	Volume I of Gergonne's *Annales*		
		1814	Fraunhofer lines
1815	"The Analytical Society" at Cambridge	1815	Battle of Waterloo
1817	Bolzano: *Rein analytischer Beweis*	1817	Optical transverse vibrations (Young and Fresnel)
1820	Oersted discovered electromagnetism		
1822	Poncelet: *Traité*; Fourier series; Feuerbach's theorem		
1826	Crelle's *Journal* founded	1826	Ampère's work in electrodynamics
	Principle of Duality (Poncelet, Plücker, Gergonne)		
	Elliptic functions (Abel, Gauss, Jacobi)		
1827	Homogeneous coordinates (Möbius, Plücker, Feuerbach)	1827	Ohm's law
	Cauchy: *Calculus of Residues*		
1828	Green: *Electricity and Magnetism*	1828	Synthesis of urea by Wöhler
1829	Lobachevskian geometry	1829	Death of Thomas Young
	Death of Abel at age 26		
1830	Peacock: *Algebra*	1830	Lyell: *Principles of Geology*
			Comte: *Cours de philosophie positive*
		1831	Faraday's electromagnetic induction
1832	Bolyai: *Absolute Science of Space*	1832	Babbage's analytical engine
	Death of Galois at age 20		
1834	Steiner became professor at Berlin		
1836	Liouville's *Journal* founded	1836	First telegraph
1837	*Cambridge and Dublin Mathematical Journal*		
		1842	Conservation of energy (Mayer and Joule)
1843	Hamilton's quaternions		
1844	Grassmann: *Ausdehnungslehre*		
		1846	Discovery of Neptune (Adams and Leverrier)
			Use of anesthesia
1847	Von Staudt: *Geometrie der Lage*		
		1848	Marx: *Communist Manifesto*
		1850	Dickens: *David Copperfield*

1852	Chasles: *Traité de géométrie supérieure*		
1854	Riemann's Habilitationschrift		
	Boole: *Laws of Thought*		
1855	Dirichlet succeeded Gauss at Göttingen		
		1858	Atlantic cable
		1859	Darwin: *Origin of Species*
			Chemical spectroscopy (Bunsen and Kirchhoff)
1863	Cayley appointed at Cambridge		
1864	Weierstrass appointed at Berlin		
		1868	Cro-Magnon caves discovered
		1869	Opening of Suez Canal
			Mendeleef's periodic table
1872	Dedekind: *Stetigkeit und irrationale Zahlen*		
	Heine: *Elemente*		
	Méray: *Nouveau preçis*		
	Klein's Erlanger Programm		
1873	Hermite proved e transcendental	1873	Maxwell: *Electricity and Magnetism*
1874	Cantor's *Mengenlehre*		
		1876	Bell's telephone
1877	Sylvester appointed at Johns Hopkins		
1881	Gibbs: *Vector Analysis*		
1882	Lindmann proved π transcendental		
1884	Frege: *Grundlagen der Arithmetik*		
		1887	Discovery of hertzian waves
1888	Beginnings of American Mathematical Society	1888	Pasteur Institute founded
1889	Peano's axioms		
1895	Poincaré: *Analysis situs*	1895	Discovery of X-rays (Roentgen)
1896	Prime number theorem proved (Hadamard and De la Vallée-Poussin)	1896	Discovery of radioactivity (Becquerel)
		1897	Discovery of electron (J. J. Thomson)
		1898	Discovery of radium (Marie Curie)
1899	Hilbert: *Grundlagen der Geometrie*		
1900	Hilbert's problems	1900	Freud: *Die Traumdeutung*
	Volume I of Russell and Whitehead: *Principia*		
		1901	Planck's quantum theory
1903	Lebesgue integration	1903	First powered air flight
		1905	Special relativity (Einstein)
1906	Functional calculus (Fréchet)		
1907	Brouwer and intuitionism		
1914	Hausdorff: *Grundzüge der Mengenlehre*	1914	Assassination of Austrian Archduke
		1915	Panama Canal opened
1916	Einstein's general theory of relativity		
1917	Hardy and Ramanujan on theory of numbers	1917	Russian Revolution

1923	Banach spaces		
		1927	Lindbergh flew the Atlantic
		1928	Fleming discovered penicillin
1930	Weyl succeeded Hilbert at Göttingen		
1931	Gödel's theorem		
1933	Weyl resigned at Göttingen	1933	Hitler became chancellor
1934	Gelfond's theorem		
1939	Volume I of Bourbaki: *Eléments*		
		1941	Pearl Harbor
		1945	Bombing of Hiroshima
		1946	First meeting of U.N.
1955	Homological algebra (Cartan and Eilenberg)		
1963	Paul J. Cohen on continuum hypothesis	1963	Assassination of President Kennedy
		1965	Death of Sir Winston Churchill
1966	15th International Congress of Mathematicians (Moscow)		

Republic, Book VII

Plato

What sort of knowledge is there, Glaucon, which would draw the soul from becoming to being? And I have in mind another consideration: You will remember that our young men are to be warrior athletes?

Yes, that was said.

Then this new kind of knowledge must have an additional quality?

From *The Dialogues of Plato*, trans. Benjamin Jowett, 4th ed., 1953, Vol. II. Reprinted by permission of the Clarendon Press, Oxford.

What quality?
It should not be useless to warriors.
Yes, if possible.
There were two parts in our former scheme of education, were there not?
Just so.
There was gymnastic which presided over the growth and decay of the body, and may therefore be regarded as having to do with generation and corruption?
True.
Then that is not the knowledge which we are seeking to discover?
No.
But what do you say of music, to the same extent as in our former scheme?
Music, he said, as you will remember, was the counterpart of gymnastic, and trained the guardians by the influences of habit, by harmony making them harmonious, by rhythm rhythmical, but not giving them science; and the words, whether fabulous or closer to the truth, were meant to impress upon them habits similar to these. But in music there was nothing which tended to that good which you are now seeking.
You are most accurate, I said, in your reminder; in music there certainly was nothing of the kind. But what branch of knowledge is there, my dear Glaucon, which is of the desired nature; since all the useful arts were reckoned mean by us?
Undoubtedly; and yet what study remains, distinct both from music and gymnastic and from the arts?
Well, I said, if nothing remains outside them, let us select something which is a common factor in all.
What may that be?
Something, for instance, which all arts and sciences and intelligences use in common, and which everyone has to learn among the first elements of education.
What is that?
The little matter of distinguishing one, two, and three—in a word, number and calculation:—do not all arts and sciences necessarily partake of them?
Yes.
Then the art of war partakes of them?
To be sure.
Then Palamedes, whenever he appears in tragedy, proves Agamemnon ridiculously unfit to be a general. Did you never remark how he declares that he had invented number, and had measured out the camping-ground at Troy, and numbered the ships and everything else; which implies that they had never been numbered before, and Agamemnon must be supposed literally to have been incapable of counting his own feet—how could he if he was ignorant of number? And if that is true, what sort of general must he have been?

I should say a very strange one, if this was as you say.

Can we deny that a warrior should have a knowledge of arithmetic?

Certainly he should, if he is to have the smallest understanding of military formations, or indeed, I should rather say, if he is to be a man at all.

I should like to know whether you have the same notion which I have of this study?

What is your notion?

It appears to me to be a study of the kind which we are seeking, and which leads naturally to reflection, but never to have been rightly used; for it has a strong tendency to draw the soul towards being.

How so? he said.

I will try to explain my meaning, I said; and I wish you would share the inquiry with me, and say 'yes' or 'no' when I attempt to distinguish in my own mind what branches of knowledge have this attracting power, in order that we may have clearer proof that arithmetic is, as I suspect, one of them.

Explain, he said.

Do you follow me when I say that objects of sense are of two kinds? some of them do not invite the intelligence to further inquiry because the sense is an adequate judge of them; while in the case of other objects sense is so untrustworthy that inquiry by the mind is imperatively demanded.

You are clearly referring, he said, to the appearance of objects at a distance, and to painting in light and shade.

No, I said, you have not quite caught my meaning.

Then what things do you mean?

When speaking of uninviting objects, I mean those which do not pass straight from one sensation to the opposite; inviting objects are those which do; in this latter case the sense coming upon the object, whether at a distance or near, does not give one particular impression more strongly than its opposite. An illustration will make my meaning clearer:—here are three fingers—a little finger, a second finger, and a middle finger.

Very good.

You may suppose that they are seen quite close: And here comes the point.

What is it?

Each of them equally appears a finger, and in this respect it makes no difference whether it is seen in the middle or at the extremity, whether white or black, or thick or thin, or anything of that kind. In these cases a man is not compelled to ask of thought the question what is a finger? for the sight never intimates to the mind that a finger is the opposite of a finger.

True.

And therefore, I said, there is nothing here which is likely to invite or excite intelligence.

There is not, he said.

But is this equally true of the greatness and smallness of the fingers? Can sight adequately perceive them? and is no difference made by the circumstance that one of the fingers is in the middle and another at the extremity? And in like manner does the touch adequately perceive the qualities of thickness or thinness, or softness or hardness? And so of the other senses; do they give perfect intimations of such matters? Is not their mode of operation on this wise—the sense which is concerned with the quality of hardness is necessarily concerned also with the quality of softness, and only intimates to the soul that the same thing is felt to be both hard and soft?

It is, he said.

And must not the soul be perplexed at this intimation which this sense gives of a hard which is also soft? What, again, is the meaning of light and heavy, if the sense pronounces that which is light to be also heavy, and that which is heavy, light?

Yes, he said, these intimations which the soul receives are very curious and require to be explained.

Yes, I said, and in these perplexities the soul naturally summons to her aid calculation and intelligence, that she may see whether the several objects announced to her are one or two.

True.

And if they turn out to be two, is not each of them one and different?

Certainly.

And if each is one, and both are two, she will conceive the two as in a state of division, for if they were undivided they could only be conceived of as one?

True.

The eye, also, certainly did see both small and great, but only in a confused manner; they were not distinguished.

Yes.

Whereas on the contrary the thinking mind, intending to light up the chaos, was compelled to reconsider the small and great viewing them as separate and not in that confusion.

Very true.

Is it not in some such way that there arises in our minds the inquiry 'What is great?' and 'What is small?'

Exactly so.

And accordingly we made the distinction of the visible and the intelligible.

A very proper one.

This was what I meant just now when I spoke of impressions which invited the intellect, or the reverse—those which strike our sense simultaneously with opposite impressions, invite thought; those which are not simultaneous with them, do not awaken it.

I understand now, he said, and agree with you.

And to which class do unity and number belong?

I do not know, he replied.

Think a little and you will see that what has preceded will supply the answer; for if simple unity could be adequately perceived by the sight or by any other sense, then, as we were saying in the case of the finger, there would be nothing to attract towards being; but when something contrary to unity is always seen at the same time, so that there seems to be no more reason for calling it one than the opposite, some discriminating power becomes necessary, and in such a case the soul in perplexity, is obliged to rouse her power of thought and to ask: "What *is* absolute unity?" This is the way in which the study of the one has a power of drawing and converting the mind to the contemplation of true being.

And surely, he said, this occurs notably in the visual perception of unity; for we see the same thing at once as one and as infinite in multitude?

Yes, I said; and this being true of one must be equally true of all number?

Certainly.

And all arithmetic and calculation have to do with number?

Yes.

And they appear to lead the mind towards truth?

Yes, in a very remarkable manner.

Then this is a discipline of the kind for which we are seeking; for the man of war must learn the art of number or he will not know how to array his troops, and the philosopher also, because he has to rise out of the sea of change and lay hold of true being, or be for ever unable to calculate and reason.

That is true.

But our guardian is, in fact, both warrior and philosopher?

Certainly.

Then this is a kind of knowledge which legislation may fitly prescribe; and we must endeavour to persuade those who are to be the principal men of our State to go and learn arithmetic, and take up the study in no amateurish spirit but pursue it until they can view the nature of numbers with the unaided mind; nor again, like merchants or retail-traders, with a view to buying or selling, but for the sake of their military use, and of the soul herself, because this will be the easiest way for her to pass from becoming to truth and being.

That is excellent, he said.

Yes, I said, and now having spoken of it, I must add how charming the science is! and in how many ways it conduces to our desired end, if pursued in the spirit of a philosopher, and not of a shopkeeper!

How do you mean?

I mean that arithmetic has, in a marked degree, that elevating effect of which we were speaking, compelling the soul to reason about abstract number, and rebelling against the introduction of numbers which have visible or tangible bodies into the argument. You know how steadily the masters of the art repel and ridicule anyone who attempts to divide the perfect unit when he is calculating, and if you divide, they multiply,* taking care that the unit shall continue one and not appear to break up into fractions.

That is very true.

Now, suppose a person were to say to them: O my friends, what are these wonderful numbers about which you are reasoning, in which, as you say, there is a unity such as you demand, and each unit is equal, invariable, indivisible,—what would they answer?

They would answer, as I should conceive, that they were speaking of those numbers which can only be grasped by thought, and not handled in any other way.

Then you see that this study may be truly called necessary for our purpose, since it evidently compels the soul to use the pure intelligence in the attainment of pure truth?

Yes; that is a marked characteristic of it.

And have you further observed, that those who have a natural talent for calculation are generally quick at every other kind of study; and even the dull, if they have been trained and exercised in this, although they may derive no other advantage from it, always become much quicker than they would otherwise have been?

Very true, he said.

And indeed, you will not easily find a study of which the learning and exercise require more pains, and not many which require as much.

You will not.

And, for all these reasons, arithmetic is a kind of knowledge in which the best natures should be trained, and which must not be given up.

I agree.

Let this then be adopted as one of our subjects of education. And next, shall we inquire whether the kindred science also concerns us?

You mean geometry?

Exactly so.

Clearly, he said, we are concerned with that part of geometry which relates to war; for in pitching a camp, or taking up a position, or closing or extending the lines

---

* Meaning either (1) that they integrate the number because they deny the possibility of fractions; or (2) that division is regarded by them as a process of multiplication, for the fractions of one continue to be units.

of an army, or any other military manoeuvre, whether in actual battle or on a march, it will make all the difference whether a general is or is not a geometrician.

Yes, I said, but for that purpose a very little of either geometry or calculation will be enough; the question relates rather to the greater and more advanced part of geometry—whether that tends in any degree to make more easy the vision of the Idea of good; and thither, as I was saying, all things tend which compel the soul to turn her gaze towards that place where is the full perfection of being, which she ought, by all means, to behold.

True, he said.

Then if geometry compels us to view being, it concerns us; if becoming only, it does not concern us?

Yes, that is what we assert.

Yet anybody who has the least acquaintance with geometry will not deny that such a conception of the science is in flat contradiction to the ordinary language of geometricians.

How so?

They speak, as you doubtless know, in terms redolent of the workshop. As if they were engaged in action, and had no other aim in view in all their reasoning, they talk of squaring, applying, extending and the like, whereas, I presume, the real object of the whole science is knowledge.

Certainly, he said.

Then must not a further admission be made?

What admission?

That the knowledge at which geometry aims is knowledge of eternal being, and not of aught which at a particular time comes into being and perishes.

That, he replied, may be readily allowed, and is true.

Then, my noble friend, geometry will draw the soul towards truth, and create the spirit of philosophy, and raise up that which is now unhappily allowed to fall down.

Nothing will be more likely to have such an effect.

Then nothing should be more sternly laid down than that the inhabitants of your fair city should by no means remain unversed in geometry. Moreover the science has indirect effects, which are not small.

Of what kind? he said.

There are the military advantages of which you spoke, I said; and further, we know that for the better apprehension of any branch of knowledge, it makes all the difference whether a man has a grasp of geometry or not.

Yes indeed, he said, all the difference in the world.

Then shall we propose this as a second branch of knowledge which our youth will study?

Let us do so, he replied.

And suppose we make astronomy the third—what do you say?

I am strongly inclined to it, he said; the observation of the seasons and of months and years is as essential to the general as it is to the farmer or sailor.

I am amused, I said, at your fear of the world, lest you should appear as an ordainer of useless studies; and I quite admit that it is by no means easy to believe that in every man there is an eye of the soul which, when by other pursuits lost and dimmed, is purified and reillumined by these studies; and is more precious far than ten thousand bodily eyes, for by it alone is truth seen. Now there are two classes of persons: some who will agree with you and will take your words as a revelation; another class who have never perceived this truth will probably find them unmeaning, for they see no noticeable profit which is to be obtained from them. And therefore you had better decide at once with which of the two you are proposing to argue. You will very likely say with neither, and that your chief aim in carrying on the argument is your own improvement, while at the same time you would not grudge to others any benefit which they may receive.

I should prefer, he said, to speak and inquire and answer mainly on my own behalf.

Then take a step backward, for we have gone wrong in the order of the sciences.

What was the mistake? he said.

After plane geometry, I said, we proceeded at once to solids in revolution, instead of taking solids in themselves; whereas after the second dimension the third, which is concerned with cubes and dimensions of depth, ought to have followed.

That is true, Socrates; but so little seems to have been discovered as yet about these subjects.

Why, yes, I said, and for two reasons:—in the first place, no government patronizes them; this leads to a want of energy in the pursuit of them, and they are difficult; in the second place, students cannot learn them unless they have a director. But then a director can hardly be found, and even if he could, as matters now stand, the students, who are very conceited, would not attend to him. That, however, would be otherwise if the whole State were to assist the director of these studies by giving honour to them; then disciples would show obedience,* and there would be continuous and earnest search, and discoveries would be made; since even now, disregarded as they are by the world, and maimed of their fair proportions, because those engaged in the research have no conception of its use, still these studies force their way by their natural charm, and it would not be surprising if they should some day emerge into light.†

* [Or, "be persuaded of the importance of the study."]

† [Or, "if the problems should be solved."]

Yes, he said, there is a remarkable charm in them. But I do not clearly understand the change in the order. By geometry, I suppose that you meant the theory of plane surfaces?

Yes, I said.

And you placed astronomy next, and then you made a step backward?

Yes, and my haste to cover the whole field has made me less speedy; the ludicrous state of research in solid geometry, which, in natural order, should have followed, made me pass over this branch and go on to astronomy, or motion of solids.

True, he said.

Then assuming that the science now omitted would come into existence if encouraged by the State, let us take astronomy as our fourth study.

Republic, Book X

Plato

Can you give me a general definition of imitation? for I really do not myself understand what it professes to be.

A likely thing, then, that I should know.

There would be nothing strange in that, for the duller eye may often see a thing sooner than the keener.

Very true, he said; but in your presence, even if I had any faint notion, I could not muster courage to utter it. Will you inquire yourself?

Well then, shall we begin the inquiry at this point, following our usual method: Whenever a number of individuals have a common name, we assume that there is one

From *The Dialogues of Plato*, trans. Benjamin Jowett, 4th ed., 1953, Vol. II. Reprinted by permission of the Clarendon Press, Oxford.

corresponding idea or form*:—do you understand me?

I do.

Let us take, for our present purpose, any instance of such a group; there are beds and tables in the world—many of each, are there not?

Yes.

But there are only two ideas or forms of such furniture—one the idea of a bed, the other of a table.

True.

And the maker of either of them makes a bed or he makes a table for our use, in accordance with the idea—that is our way of speaking in this and similar instances—but no artificer makes the idea itself: how could he?

Impossible.

And there is another artificer,—I should like to know what you would say of him.

Who is he?

One who is the maker of all the works of all other workmen.

What an extraordinary man!

Wait a little, and there will be more reason for your saying so. For this is the craftsman who is able to make not only furniture of every kind, but all that grows out of the earth, and all living creatures, himself included; and besides these he can make earth and sky and the gods, and all the things which are in heaven or in the realm of Hades under the earth.

He must be a wizard and no mistake.

Oh! you are incredulous, are you? Do you mean that there is no such maker or creator, or that in one sense there might be a maker of all these things but in another not? Do you see that there is a way in which you could make them all yourself?

And what way is this? he asked.

An easy way enough; or rather, there are many ways in which the feat might be quickly and easily accomplished, none quicker than that of turning a mirror round and round—you would soon enough make the sun and the heavens, and the earth and yourself, and other animals and plants, and furniture and all the other things of which we were just now speaking, in the mirror.

Yes, he said; but they would be appearances only.

Very good, I said, you are coming to the point now. And the painter too is, as I conceive, just such another—a creator of appearances, is he not?

Of course.

---

* [Or, (probably better): "we have been accustomed to assume that there is one single idea corresponding to each group of particulars; and to these we give the same name (as we give the idea)."]

But then I suppose you will say that what he creates is untrue. And yet there is a sense in which the painter also creates a bed? Is there not?

Yes, he said, but here again, an appearance only.

And what of the maker of the bed? were you not saying that he too makes, not the idea which according to our view is the real object denoted by the word bed, but only a particular bed?

Yes, I did.

Then if he does not make a real object he cannot make what *is*, but only some semblance, of existence; and if any one were to say that the work of the maker of the bed, or of any other workman, has real existence, he could hardly be supposed to be speaking the truth.

Not, at least, he replied, in the view of those who make a business of these discussions.

No wonder, then, that his work too is an indistinct expression of truth.

No wonder.

Suppose now that by the light of the examples just offered we inquire who this imitator is?

If you please.

Well then, here we find three beds: one existing in nature, which is made by God, as I think that we may say—for no one else can be the maker?

No one, I think.

There is another which is the work of the carpenter?

Yes.

And the work of the painter is a third?

Yes.

Beds, then, are of three kinds, and there are three artists who superintend them: God, the maker of the bed, and the painter?

Yes, there are three of them.

God, whether from the choice or from necessity, made one bed in nature and one only; two or more such beds neither ever have been nor ever will be made by God.

Why is that?

Because even if He had made but two, a third would still appear behind them of which they again both possessed the form, and that would be the real bed and not the two others.

Very true, he said.

God knew this, I suppose, and He desired to be the real maker of a real bed, not a kind of maker of a kind of bed, and therefore He created a bed which is essentially and by nature one only.

So it seems.

Shall we, then, speak of Him as the natural author or maker of the bed?

Yes, he replied; inasmuch as by the natural process of creation He is the author of this and of all other things.

And what shall we say of the carpenter—is not he also the maker of a bed?

Yes.

But would you call the painter an artificer and maker?

Certainly not.

Yet if he is not the maker, what is he in relation to the bed?

I think, he said, that we may fairly designate him as the imitator of that which the others make.

Good, I said; then you call him whose product is third in the descent from nature, an imitator?

Certainly, he said.

And so if the tragic poet is an imitator, he too is thrice removed from the king and from the truth; and so are all other imitators.

That appears to be so.

Then about the imitator we are agreed. And what about the painter?—Do you think he tries to imitate in each case that which originally exists in nature, or only the creations of artificers?

The latter.

As they are or as they appear? you have still to determine this.

What do you mean?

I mean to ask whether a bed really becomes different when it is seen from different points of view, obliquely or directly or from any other point of view? Or does it simply appear different, without being really so? And the same of all things.

Yes, he said, the difference is only apparent.

Now let me ask you another question: Which is the art of painting designed to be—an imitation of things as they are, or as they appear—of appearance or of reality?

Of appearance, he said.

Then the imitator is a long way off the truth, and can reproduce all things because he lightly touches on a small part of them, and that part an image. For example: A painter will paint a cobbler, carpenter, or any other artisan, though he knows nothing of their arts; and, if he is a good painter, he may deceive children or simple persons when he shows them his picture of a carpenter from a distance, and they will fancy that they are looking at a real carpenter.

Certainly.

And surely, my friend, this is how we should regard all such claims: whenever any one informs us that he has found a man who knows all the arts, and all things else that anybody knows, and every single thing with a higher degree of accuracy than any other man—whoever tells us this, I think that we can only retort that he is a simple creature who seems to have been deceived by some wizard or imitator whom he met, and whom he thought all-knowing, because he himself was unable to analyse the nature of knowledge and ignorance and imitation.

Most true.

And next, I said, we have to consider tragedy and its leader, Homer; for we hear some persons saying that these poets know all the arts; and all things human; where virtue and vice are concerned, and indeed all divine things too; because the good poet cannot compose well unless he knows his subject, and he who has not this knowledge can never be a poet. We ought to consider whether here also there may not be a similar illusion. Perhaps they may have come across imitators and been deceived by them; they may not have remembered when they saw their works that these were thrice removed from the truth, and could easily be made without any knowledge of the truth, because they are appearances only and not realities? Or, after all, they may be in the right, and good poets do really know the things about which they seem to the many to speak so well?

The question, he said, should by all means be considered.

Now do you suppose that if a person were able to make the original as well as the image, he would seriously devote himself to the image-making branch? Would he allow imitation to be the ruling principle of his life, as if he had nothing higher in him?

I should say not.

But the real artist, who had real knowledge of those things which he chose also to imitate, would be interested in realities and not in imitations; and would desire to leave as memorials of himself works many and fair; and, instead of being the author of encomiums, he would prefer to be the theme of them . . . .

But we have not yet brought forward the heaviest count in our accusation:—the power which poetry has of harming even the good (and there are very few who are not harmed), is surely an awful thing?

Yes, certainly, if the effect is what you say.

Hear and judge: The best of us, as I conceive, when we listen to a passage of Homer or one of the tragedians, in which he represents some hero who is drawling out his sorrows in a long oration, or singing, and smiting his breast—the best of us, you know, delight in giving way to sympathy, and are in raptures at the excellence of the poet who stirs our feelings most.

Yes, of course I know.

But when any sorrow of our own happens to us, then you may observe that we pride ourselves on the opposite quality—we would fain be quiet and patient; this is considered the manly part, and the other which delighted us in the recitation is now deemed to be the part of a woman.

Very true, he said.

Now can we be right in praising and admiring another who is doing that which any one of us would abominate and be ashamed of in his own person?

No, he said, that is certainly not reasonable.

Nay, I said, quite reasonable from one point of view.

What point of view?

If you consider, I said, that when in misfortune we feel a natural hunger and desire to relieve our sorrow by weeping and lamentation, and that this very feeling which is starved and suppressed in our own calamities is satisfied and delighted by the poets;—the better nature in each of us, not having been sufficiently trained by reason or habit, allows the sympathetic element to break loose because the sorrow is another's; and the spectator fancies that there can be no disgrace to himself in praising and pitying any one who while professing to be a brave man, gives way to untimely lamentation; he thinks that the pleasure is a gain, and is far from wishing to lose it by rejection of the whole poem. Few persons ever reflect, as I should imagine, that the contagion must pass from others to themselves. For the pity which has been nourished and strengthened in the misfortunes of others is with difficulty repressed in our own.

How very true!

And does not the same hold also of the ridiculous? There are jests which you would be ashamed to make yourself, and yet on the comic stage, or indeed in private, when you hear them, you are greatly amused by them, and are not at all disgusted at their unseemliness;—the case of pity is repeated;—there is a principle in human nature which is disposed to raise a laugh, and this, which you once restrained by reason because you were afraid of being thought a buffoon, is now let out again; and having stimulated the risible faculty at the theatre, you are betrayed unconsciously to yourself into playing the comic poet at home.

Quite true, he said.

And the same may be said of lust and anger and all the other affections, of desire and pain and pleasure, which are held to be inseparable from every action—in all of them poetry has a like effect; it feeds and waters the passions instead of drying them up; she lets them rule, although they ought to be controlled if mankind are ever to increase in happiness and virtue.

I cannot deny it.

Therefore, Glaucon, I said, whenever you meet with any of the eulogists of Homer declaring that he has been the educator of Hellas, and that he is profitable for education and for the ordering of human things, and that you should take him up again and again and get to know him and regulate your whole life according to him, we may love and honour those who say these things—they are excellent people, as far as their lights extend; and we are ready to acknowledge that Homer is the greatest of poets and first of tragedy writers; but we must remain firm in our conviction that hymns to the gods and praises of famous men are the only poetry which ought to be admitted into our State. For if you go beyond this and allow the honeyed Muse to enter, either in epic or lyric verse, not law and the reason of mankind, which by common consent have ever been deemed best,* but pleasure and pain will be the rulers in our State.

Questions on Books VII and X

1. How does Plato define mathematics and its uses?
2. Of what two kinds are the objects of sense perception and how does Plato's notion of mathematics relate to them?
3. What does Plato mean when he uses the word "soul"?
4. How are the concepts of unity and infinity similar?
5. Defend or argue against the notions that unity and infinity can exist only in the mind. Reverse your position.
6. Distinguish between nature and nurture.
7. What, for Plato, is the knowledge at which mathematics aims?
8. Why does Plato say one should study mathematics?
9. Why does Plato regard mathematics as a science? Do you think mathematics is a science? If so, why? If not, why not?
10. Use Book X to find Plato's definition of an artist. Compare Plato and Halmos on the mathematician as an artist.
11. Look up the philosophical meanings of being and becoming.

---

*[Or: "law, and the principle which the community in every case has pronounced to be the best."]

Part Three

Potpourri

Introduction

In mathematics, definitions must always be precise. The next two articles deal explicitly with the problems this raises.

The first article, by Immanuel Kant, is from one of his most frequently read but least understood works. Here he deals with one of the oldest problems considered by the Greeks: whether various questions can be answered by pure reason, or whether facts perceived about the universe need to be brought into the picture. Propositions whose truth (or falsity) can be decided by pure reason, prior to experience, are called *a priori*. Propositions whose truth value can be determined only after facts are introduced are called *a posteriori*.

In real life Lewis Carroll was the Reverend Charles Lutwidge Dodgson, a mathematics and logic lecturer at Oxford University in the mid-nineteenth century. Though most people know the stories that go under the general heading of *Alice in Wonderland*, they often fail to realize that there are usually two books published under the same heading. Of the two, the most appropriate for a mathematics course is *Through the Looking Glass*.

It is, therefore, from *Through the Looking Glass* that I have chosen to quote. Several things are remarkable about the story. It takes place on a chessboard as a chess game. Of course, chess has long been associated with mathematics. The opening lines of the selection reflect the "Let's pretend" attitude that, when applied to the mainstream of mathematics, is translated as "What if" or simply "If—be the case."

Through the Looking Glass also contains a contrasting view of definition to that found in Kant. H. Dumpty offers us a splendid discourse on this matter. This discourse, however, should be coupled with those of W. Knight when Alice confuses word with object. If one puts all three ideas together, one has a good beginning course in definition theory.

Critique of Pure Reason

Immanuel Kant

I. *Definitions.*—To *define*, as the word itself indicates, really only means to present the complete, original concept of a thing within the limits of its concept. (*Completeness* means clearness and sufficiency of characteristics; by *limits* is meant the precision shown in there not being more of these characteristics than belong to the complete concept; by *original* is meant that this determination of these limits is not derived from anything else, and therefore does not require any proof; for if it did, that would disqualify the supposed explanation from standing at the head of all the judgments regarding its object.) If this be our standard, an *empirical* concept cannot be defined at all, but only made *explicit.* For since we find in it only a few characteristics of a certain species of sensible object, it is never certain that we are not using the word, in denoting one and the same object, sometimes so as to stand for more, and sometimes so as to stand for fewer characteristics. Thus in the concept of *gold* one man may think, in addition to its weight, colour, malleability, also of its property of resisting rust, while another will perhaps know nothing of this quality. We make use of certain characteristics only so long as they are adequate for the purpose of making distinctions; new observations remove some properties and add others; and thus the limits of the concept are never assured. And indeed what useful purpose could be served by defining an empirical concept, such, for example, as that of water? When we speak of water and its properties, we do not stop short at what is thought in the word, water, but proceed to experiments. The word, with the few

From *The Critique of Pure Reason*, ed. Norman Kemp Smith. St. Martin's Press, Inc., 1956. Reprinted by permission of St. Martin's Press, Inc., and Macmillan London Basingstoke.

characteristics which we attach to it, is more properly to be regarded as merely a designation than as a concept of the thing; the so-called definition is nothing more than a determining of the word. In the second place, it is also true that no concept given *a priori*, such as substance, cause, right, equity, etc., can, strictly speaking, be defined. For I can never be certain that the clear representation of a given concept, which as given may still be confused, has been completely effected, unless I know that it is adequate to its object. But since the concept of it may, as given, include many obscure representations, which we overlook in our analysis, although we are constantly making use of them in our application of the concept, the completeness of the analysis of my concept is always in doubt, and a multiplicity of suitable examples suffices only to make the completeness *probable*, never to make it *apodeictically* certain. Instead of the term, definition, I prefer to use the term, exposition, as being a more guarded term, which the critic can accept as being up to a certain point valid, though still entertaining doubts as to the completeness of the analysis. Since, then, neither empirical concepts nor concepts given *a priori* allow of definition, the only remaining kind of concepts, upon which this mental operation can be tried, are arbitrarily invented concepts. A concept which I have invented I can always define; for since it is not given to me either by the nature of understanding or by experience, but is such as I have myself deliberately made it to be, I must know what I have intended to think in using it. I cannot however say that I have thereby defined a true object. For if the concept depends on empirical condition, as, e.g., the concept of a ship's clock, this arbitrary concept of mine does not assure me of the existence or of the possibility of its object. I do not even know from it whether it has an object at all, and my explanation may better be described as a declaration of my project than as a definition of an object. There remain, therefore, no concepts which allow of definition, except only those which contain an arbitrary synthesis that admits of *a priori* construction. Consequently, mathematics is the only science that has definitions. For the object which it thinks it exhibits *a priori* is intuition, and this object certainly cannot contain either more or less than the concept, since it is through the definition that the concept of the object is given—and given originally, that is, without its being necessary to derive the definition from any other source. The German language has for the (Latin) terms *exposition, explication, declaration,* and *definition* only one word, Erklärung, and we need not, therefore, be so stringent in our requirements as altogether to refuse to philosophical explanations the honourable title, definition. We shall confine ourselves simply to remarking that while philosophical definitions are never more than expositions of given concepts, mathematical definitions are constructions of concepts, originally framed by the mind itself; and that while the former can be obtained only by analysis (the completeness of which is never apodeictically certain), the latter are produced synthetically.

Whereas, therefore, mathematical definitions *make* their concepts, in philosophical definitions concepts are only *explained*. From this it follows:

(a) That in philosophy we must not imitate mathematics by beginning with definitions, unless it be by way simply of experiment. For since the definitions are analyses of given concepts, they presuppose the prior presence of the concepts, although in a confused state; and the incomplete exposition must precede the complete. Consequently, we can infer a good deal from a few characteristics, derived from an incomplete analysis, without having yet reached the complete exposition, that is, the definition. In short, the definition in all its precision and clarity ought, in philosophy, to come rather at the end than at the beginning of our enquiries. (Philosophy is full of faulty definitions, especially of definitions which, while indeed containing some of the elements required, are yet not complete. If we could make no use of a concept till we had defined it, all philosophy would be in a pitiable plight. But since a good and safe use can still be made of the elements by analysis so far as they go, defective definitions, that is, propositions which are properly not definitions, but are yet true, and are therefore approximations to definitions, can be employed with great advantage. In mathematics definition belongs *ad esse*, in philosophy *ad melius esse*. It is desirable to attain an adequate definition, but often very difficult. The jurists are still without a definition of their concept of right.) In mathematics, on the other hand, we have no concept whatsoever prior to the definition, through which the concept itself is first given. For this reason mathematical science must always begin, and it can always begin, with the definition.

(b) That mathematical definitions can never be in error. For since the concept is first given through the definition, it includes nothing except precisely what the definition intends should be understood by it. But although nothing incorrect can be introduced into its content, there may sometimes, though rarely, be a defect in the form in which it is clothed, namely as regards precision. Thus the common explanation of the circle that it is a *curved* line every point in which is equidistant from one and the same point (the centre), had the defect that the determination, curved, is introduced unnecessarily. For there must be a particular theorem, deduced from the definition and easily capable of proof, namely, that if all points in a line are equidistant from one and the same point, the line is curved (no part of it straight). Analytic definitions, on the other hand, may err in many ways, either through introducing characteristics which do not really belong to the concept, or by lacking that completeness which is the essential feature of a definition. The latter defect is due to the fact that we can never be quite certain of the completeness of the analysis. For these reasons the mathematical method of definition does not admit of imitation in philosophy.

Questions on Kant

1. What does Kant mean by an empirical concept? Why does he say it cannot be defined?
2. What does a concept given a priori mean? Why does Kant think these undefinable?
3. Do you agree or disagree with Kant in the above position and why?
4. What concept does Kant say can be defined?
5. How does Kant view mathematics?
6. What does Kant say the nature of mathematical definition is?
7. Read *Prolegomenon to any Future Metaphysics*. Tell briefly the meaning of the following groups of words:
    (a) synthetic a priori
    (b) synthetic a posteriori
    (c) analytic a priori
    (d) analytic a posteriori
In this type of classification where does mathematics lie?

Through the Looking Glass

Lewis Carroll

Chapter I

Looking-Glass House

... "Kitty, can you play chess? Now, don't smile, my dear; I'm asking it seriously. Because, when we were playing just now, you watched just as if you understood it; and when I said "Check!" you purred! Well, it was a nice check, Kitty, and really, I might have won, if it hadn't been for that nasty Knight, that came wriggling down among my pieces. Kitty dear, let's pretend . . ." And here I wish I could tell you half the things Alice used to say, beginning with her favorite phrase, "Let's pretend." She had had quite a long argument with her sister only the day before—all because Alice had begun with, "Let's pretend we're kings and queens"; and her sister, who liked being very exact, had argued that they couldn't, because there were only two of them, and Alice had been reduced at last to say, "Well *you* can be one of them then, and *I'll* be all the rest." And once she had frightened her old nurse by shouting suddenly in her ear, "Nurse? Do let's pretend that I'm a hungry hyena, and you're a bone!"

But this is taking us away from Alice's speech to the kitten. "Let's pretend that you're the Red Queen, Kitty! Do you know, I think if you sat up and folded your arms, you'd look exactly like her. Now do try, there's a dear!" And Alice got the Red Queen off the table, and set it up before the kitten as a model for it to imitate; however, the thing didn't succeed, principally, Alice said, because the kitten wouldn't fold its arms properly. So, to punish it; she held it up to the Looking Glass that it might see how sulky it was—" And if you're not good directly," she added, "I'll put you through into Looking-glass House. How would you like *that*?"

"Now, if you'll only attend, Kitty, and not talk so much, I'll tell you all my ideas about Looking-glass House. First, there's the room you can see through the glass—that's just the same as our drawing room, only the things go the other way. I can see all of it when I get upon a chair—all but the bit just behind the fire-place. Oh, I do so wish I could see *that* bit! I want so much to know whether they've a fire in the winter; you never can tell, you know, unless our fire smokes, and then smoke comes up in that room, too—but that may be only pretense, just to make it look as if they had a fire. Well, then, the books are something like our books, only the words go the wrong way; I know that, because I've held up one of our books to the glass, and then they hold up one in the other room.

"How would you like to live in Looking-glass House, Kitty? I wonder if they'd give you milk in there? Perhaps Looking-glass milk isn't good to drink. But oh, Kitty! now we come to the passage. You can just see a little peep of the passage in Looking-glass House, if you leave the door of our drawing room side open; and it's very like our passage as far as you can see, only, you know, it may be quite different on beyond. Oh, Kitty, how nice it would be if we could only get through into Looking-glass House! I'm sure it's got, oh! such beautiful things in it. Let's pretend there's a way of getting through into it, somehow, Kitty. Let's pretend the glass has got all soft like gauze, so that we can get through. Why, it's turning into a sort of mist now, I declare! It'll be easy enough to get through—" She was up on the chimney piece while she said this, though she hardly knew how she got there. And certainly the glass *was* beginning to melt away, just like a bright silvery mist.

In another moment Alice was through the glass, and had jumped lightly down into the Looking-glass room. The very first thing she did was to look whether there was a fire in the fireplace, and she was quite pleased to find that there was a real one, blazing away as brightly as the one she had left behind. "So I shall be as warm here as I was in the old room," thought Alice; "warmer, in fact, because there'll be no one here to scold me away from the fire. Oh, what fun it'll be, when they see me through the glass in here, and can't get at me!"

Then she began looking about, and noticed that what could be seen from the old room was quite common and uninteresting, but that all the rest was as different as possible. For instance, the pictures on the wall next the fire seemed to be all alive, and the very clock on the chimney piece (you know you can only see the back of it in the Looking Glass) had got the face of a little old man, and grinned at her.

"They don't keep this room so tidy as the other," Alice thought to herself, as she noticed several of the chessmen down in the hearth among the cinders; but in another moment, with a little "Oh!" of surprise, she was down on her hands and knees watching them. The chessmen were walking about, two and two!...

There was a book lying near Alice on the table, and while she sat watching the White King (for she was still a little anxious about him, and had the ink all ready to throw over him, in case he fainted again), she turned over the leaves to find some part that she could read—"for it's all in some language I don't know," she said to herself.

This was the poem that Alice read:

Jabberwocky

'Twas brillig, and the slithy toves
    Did gyre and gimble in the wabe;
All mimsy were the borogoves
    And the mome raths outgrabe.

"Beware the Jabberwock, my son!
    The jaws that bite, the claws that catch!
Beware the Jubjub bird, and shun
    The frumious Bandersnatch!"

He took his vorpal sword in hand:
    Long time the manxome foe he sought—
So rested he by the Tumtum tree,
    And stood awhile in thought.

And as in uffish thought he stood,
    The Jabberwock, with eyes of flame,
Came whiffling through the tulgey wood,
    And burbled as it came!

One, two! One, two! And through and through
    The vorpal blade went snicker-snack!
He left it dead, and with its head
    He went galumphing back.

"And hast thou slain the Jabberwock?
    Come to my arms, my beamish boy!
O frabjous day! Callooh! Callay!"
    He chortled in his joy.

'Twas brillig, and the slithy toves
  Did gyre and gimble in the wabe;
All mimsy were the borogoves,
  And the mome raths outgrabe.

"It seems very pretty," she said when she had finished it; "but it's *rather* hard to understand!" (You see she didn't like to confess, even to herself, that she couldn't make it out at all.) "Somehow it seems to fill my head with ideas—only I don't exactly know what they are! However, *somebody* killed *something*; that's clear, at any rate—" . . .

Chapter II

The Garden of the Live Flowers

. . . "Where do you come from?" said the Red Queen. "And where are you going? Look up, speak nicely, and don't twiddle your fingers all the time."

Alice attended to all these directions, and explained, as well as she could, that she had lost her way.

"I don't know what you mean by your way," said the Queen; "all the ways about here belong to *me*—but why did you come out here at all?" she added in a kinder tone. "Curtsy while you're thinking what to say. It saves time."

Alice wondered a little at this, but she was too much in awe of the Queen to disbelieve it. I'll try it when I go home," she thought to herself, "the next time I'm a little late for dinner."

"It's time for you to answer now," the Queen said, looking at her watch; "open your mouth a *little* wider when you speak, and always say 'your Majesty.' "

"I only wanted to see what the garden was like, your Majesty—"

"That's right," said the Queen, patting her on the head, which Alice didn't like at all; "though, when you say 'garden'—*I've* seen gardens compared with which this would be a wilderness."

Alice didn't dare to argue the point, but when on: "And I thought I'd try and find my way to the top of that hill—"

"When you say 'hill,' " the Queen interrupted, "*I* could show you hills in comparison with which you'd call that a valley."

"No, I shouldn't," said Alice, surprised into contradicting her at last: "a hill *can't* be a valley, you know. That would be nonsense—"

The Red Queen shook her head. "You may call it 'nonsense,' if you like," she said, "but *I've* heard nonsense compared with which that would be as sensible as a dictionary!"

Alice curtsied again, as if she was afraid from the Queen's tone that she was a *little* offended, and they walked on in silence till they got to the top of the hill.

For some minutes Alice stood without speaking, looking out in all directions over the country—and a most curious country it was. There were a number of tiny little brooks running straight across it from side to side, and the ground between was divided up into squares by a number of little green hedges, that reached from brook to brook.

"I declare it's marked out just like a large chessboard!" Alice said at last. "There ought to be some men moving about somewhere—and so there are!" she added in a tone of delight, and her heart began to beat quick with excitement as she went on. "It's a great huge game of chess that's being played—all over the world—if this *is* the world at all, you know. Oh, what fun it is! How I wish I was one of them! I wouldn't mind being a Pawn, if only I might join—though, of course, I should like to be a Queen best."

She glanced rather shyly at the real Queen as she said this, but her companion only smiled pleasantly, and said, "That's easily managed. You can be the White Queen's Pawn, if you like, as Lily's too young to play; and you're in the Second Square to begin with; when you get to the Eighth Square you'll be a Queen—" Just at this moment, somehow or other, they began to run.

Alice never could quite make out, in thinking it over afterward, how it was that they began; all she remembers is that they were running hand in hand, and the Queen went so fast that it was all she could do to keep up with her; and still the Queen kept crying, "Faster! Faster!" but Alice felt she could not go faster, though she had no breath left to say so.

The most curious part of the thing was that the trees and other things round them never changed their places at all; however fast they went they never seemed to pass anything. "I wonder if all the things move along with us?" thought poor puzzled Alice.

And the Queen seemed to guess her thoughts, for she cried, "Faster! Don't try to talk!"

Not that Alice had any idea of doing *that*. She felt as if she would never be able to talk again, she was getting so much out of breath; and still the Queen cried,

"Faster! Faster!" and dragged her along. "Are we nearly there?" Alice managed to pant out at last.

"Nearly there?" the Queen repeated. "Why, we passed it ten minutes ago! Faster!" And they ran on for a time in silence, with the wind whistling in Alice's ears, and almost blowing her hair off her head, she fancied.

"Now! Now!" cried the Queen. "Faster! Faster!" And they went so fast that at last they seemed to skim through the air, hardly touching the ground with their feet till, suddenly, just as Alice was getting quite exhausted, they stopped, and she found herself sitting on the ground, breathless and giddy.

The Queen propped her up against a tree, and said kindly, "You may rest a little now."

Alice looked round her in great surprise. "Why, I do believe we've been under this tree the whole time! Everything's just as it was!"

"Of course it is," said the Queen. "What would you have it?"

"Well, in our country," said Alice, still panting a little, "you'd generally get to somewhere else—if you ran very fast for a long time, as we've been doing."

"A slow sort of country!" said the Queen. "Now, *here*, you see, it takes all the running *you* can do to keep in the same place. If you want to get somewhere else, you must run at least twice as fast as that!"

"I'd rather not try, please!" said Alice. "I'm quite content to stay here—only I *am* so hot and thirsty!"

"I know what *you'd* like!" the Queen said good-naturedly, taking a little box out of her pocket. "Have a biscuit?"

Alice thought it would not be civil to say "No," thought it wasn't at all what she wanted. So she took it, and ate it as well as she could; and it was very dry; and she thought she had never been so nearly choked in all her life.

"While you're refreshing yourself," said the Queen, "I'll just take the measurements." And she took a ribbon out of her pocket, marked in inches, and began measuring the ground, and sticking little pegs in here and there.

"At the end of two yards," she said, putting in a peg to mark the distance, "I shall give you your directions—have another biscuit?"

"No, thank you," said Alice; "one's quite enough!"

"Thirst quenched, I hope?" said the Queen.

Alice did not know what to say to this, but luckily the Queen did not wait for an answer, but went on. "At the end of three yards I shall repeat them—for fear of your forgetting them. At the end of *four* I shall say good-by. And at the end of five I shall go!"

She had got all the pegs put in by this time, and Alice looked on with great interest as she returned to the tree, and then began slowly walking down the row.

At the two-yard peg she faced round, and said, "A Pawn goes two squares in its first move, you know. So you'll go *very* quickly through the Third Square—by railway, I should think—and you'll find yourself in the Fourth Square in no time. Well, that square belongs to Tweedledum and Tweedledee—the Fifth is mostly water—the Sixth belongs to Humpty Dumpty—But you make no remark?"

"I—I don't know I had to make one—just then," Alice faltered out.

"You *should* have said," the Queen went on in a tone of grave reproof, " 'It's extremely kind of you to tell me all this'—however, we'll suppose it said—the Seventh Square is all forest—however, one of the Knights will show you the way—and in the Eighth Square we shall be Queens together, and it's all feasting and fun!" Alice got up and curtsied, and sat down again.

At the next peg the Queen turned again, and this time she said, "Speak in French when you can't think of the English for a thing—turn out your toes when you walk —and remember who you are!" She did not wait for Alice to curtsy this time, but walked on quickly to the next peg, where she turned for a moment to say "good-by" and then hurried on to the last.

How it happened Alice never knew, but exactly as she came to the last peg, she was gone. Whether she vanished into the air, or whether she ran quickly into the wood ("And she *can* run very fast!" thought Alice), there was no way of guessing, but she was gone, and Alice began to remember that she was a Pawn, and that it would soon be time for her to move . . . .

Chapter IV

Tweedledum and Tweedledee

They were standing under a tree, each with an arm around the other's neck, and Alice knew which was which in a moment, because one of them had "DUM" embroidered on his collar, and the other "DEE." "I suppose they've each got 'TWEEDLE' round at the back of the collar," she said to herself.

They stood so still that she quite forgot they were alive, and she was just looking round to see if the word "TWEEDLE" was written at the back of each collar when she was startled by a voice coming from the one marked "DUM."

"If you think we're waxworks," he said, "you ought to pay, you know. Waxworks weren't made to be looked at for nothing. Nohow!"

"Contrariwise," added the one marked "DEE," if you think we're alive, you ought to speak."

"I'm sure I'm very sorry" was all Alice could say, for the words of the old song kept ringing through her head like the ticking of a clock, and she could hardly help saying them out loud:

> "Tweedledum and Tweedledee
>   Agreed to have a battle;
> For Tweedledum said Tweedledee
>   Had spoiled his nice new rattle.
>
> Just then flew down a monstrous crow,
>   As black as a tar barrel,
> Which frightened both the heroes so,
>   They quite forgot their quarrel."

"I know what you're thinking about," said Tweedledum; "but it isn't so, nohow."

"Contrariwise," continued Tweedledee, "if it was so, it might be; and if it were so, it would be; but as it isn't, it ain't. That's logic." . . .

Chapter V

Wool and Water

. . . "I can't believe that!" said Alice.

"Can't you?" the Queen said in a pitying tone. "Try again; draw a long breath, and shut your eyes."

Alice laughed. "There's no use trying," she said; "one *can't* believe impossible things."

"I dare say you haven't had much practice," said the Queen. "When I was your age I always did it for half an hour a day. Why, sometimes I've believed as many as six impossible things before breakfast." . . .

Chapter VI

Humpty Dumpty

However, the egg only got larger and larger, and more and more human; when she had come within a few yards of it she saw that it had eyes and a nose and mouth; and when she had come close to it she saw clearly that it was HUMPTY DUMPTY himself. "It can't be anybody else!" she said to herself. "I'm as certain of it as if his name were written all over his face!"

It might have been written a hundred times, easily, on that enormous face. Humpty Dumpty was sitting with his legs crossed, like a Turk, on the top of a high wall—such a narrow one that Alice quite wondered how he could keep his balance —and, as his eyes were steadily fixed in the opposite direction, and he didn't take the least notice of her, she thought he must be a stuffed figure after all.

"And how exactly like an egg he is!" she said aloud, standing with her hands ready to catch him, for she was every moment expecting him to fall.

"It's very provoking," Humpty Dumpty said after a long silence, looking away from Alice as he spoke, "to be called an egg—very!"

"I said you *looked* like an egg, sir," Alice gently explained. "And some eggs are very pretty, you know," she added, hoping to turn her remark into a sort of compliment.

"Some people," said Humpty Dumpty, looking away from her, as usual, "have no more sense than a baby!"

Alice didn't know what to say to this; it wasn't at all like conversation, she thought, as he never said anything to her; in fact, his last remark was evidently addressed to a tree—so she stood and softly repeated to herself:

> "Humpty Dumpty sat on a wall;
> Humpty Dumpty had a great fall.
> All the King's horses and all the King's men
> Couldn't put Humpty Dumpty in his place again."

"That last line is much too long for the poetry," she added, almost out loud, forgetting that Humpty Dumpty would hear her.

"Don't stand chattering to yourself like that," Humpty Dumpty said, looking at her for the first time, "but tell me your name and your business."

"My *name* is Alice, but—"

"It's a stupid name enough!" Humpty Dumpty interrupted impatiently. "What does it mean?"

"*Must* a name mean something?" Alice asked doubtfully.

"Of course it must," Humpty Dumpty said with a short laugh. "My name means the shape I am—and a good, handsome shape it is, too. With a name like yours, you might be any shape almost."

"Why do you sit out here all alone?" said Alice, not wishing to begin an argument.

"Why, because there's nobody with me!" cried Humpty Dumpty. "Did you think I didn't know the answer to *that?* Ask another."

"Don't you think you'd be safer down on the ground?" Alice went on, not with any idea of making another riddle, but simply in her good-natured anxiety for the queer creature. "That wall is so *very* narrow!"

"What tremendously easy riddles you ask!" Humpty Dumpty growled out. "Of course I don't think so! Why, if ever I did fall off—which there's no chance of— but if I did—" Here he pursed up his lips, and looked so solemn and grand that Alice could hardly help laughing. "If I did fall," he went on, "*the King has promised me—* Ah, you may turn pale, if you like! You didn't think I was going to say that, did you? *The King has promised me—with his very own mouth—to—to—*"

"To send all his horses and all his men," Alice interrupted rather unwisely.

"Now I declare, that's too bad!" Humpty Dumpty cried, breaking into a sudden passion. "You've been listening at doors—and behind trees—and down chimneys—or you couldn't have known it!"

"I haven't, indeed!" Alice said very gently. "It's in a book."

"Ah, well! They may write such things in a book," Humpty Dumpty said in a calmer tone. "That's what you call a History of England, that is. Now, take a good look at me! I'm one that has spoken to a King, I am; mayhap you'll never see such another; and to show you I'm not proud, you may shake hands with me!" And he grinned almost from ear to ear, as he leaned forward (and as nearly as possible fell off the wall in doing so) and offered Alice his hand. She watched him a little anxiously as she took it. "If he smiled much more, the ends of his mouth might meet behind," she thought, "and then I don't know what would happen to his head! I'm afraid it would come off!"

"Yes, all his horses and all his men," Humpty Dumpty went on. "They'd pick me up again in a minute, they would! However, this conversation is going on a little too fast; let's go back to the last remark but one."

"I'm afraid I can't quite remember it," Alice said very politely.

"In that case we start fresh," said Humpty Dumpty, "and it's my turn to choose a subject." ("He talks about it just as if it was a game!" thought Alice.) "So here's a question for you. How old did you say you were?"

Alice made a short calculation, and said, "Seven years and six months."

"Wrong!" Humpty Dumpty exclaimed triumphantly. "You never said a word like it!"

"I thought you meant 'How old *are* you?'" Alice explained.

"If I'd meant that, I'd have said it," said Humpty Dumpty.

Alice didn't want to begin another argument, so she said nothing.

"Seven years and six months!" Humpty Dumpty repeated thoughtfully. "An uncomfortable sort of age. Now, if you'd asked *my* advice, I'd have said, 'Leave off at seven'; but it's too late now."

"I never ask advice about growing," Alice said indignantly.

"Too proud?" the other inquired.

Alice felt even more indignant at this suggestion. "I mean," she said, "that one can't help growing older."

"*One* can't perhaps," said Humpty Dumpty, "but *two* can. With proper assistance, you might have left off at seven."

"What a beautiful belt you've got on!" Alice suddenly remarked. (They had had quite enough of the subject of age, she thought; and if they really were to take turns in choosing subjects, it was her turn now.) "At least," she corrected herself on second thoughts, "a beautiful cravat, I should have said—no, a belt, I mean—I beg your pardon!" she added in dismay, for Humpty Dumpty looked thoroughly offended, and she began to wish she hadn't chosen that subject. "If I only knew," she thought to herself, "which was neck and which was waist!"

Evidently Humpty Dumpty was very angry, though he said nothing for a minute or two. When he *did* speak again, it was in a deep growl.

"It is a—*most—provoking*—thing," he said at last, "when a person doesn't know a cravat from a belt!"

"I know it's very ignorant of me," Alice said, in so humble a tone that Humpty Dumpty relented.

"It's a cravat, child, and a beautiful one, as you say. It's a present from the White King and Queen. There now!"

"Is it really?" said Alice, quite pleased to find that she *had* chosen a good subject, after all.

"They gave it me," Humpty Dumpty continued thoughtfully, as he crossed one knee over the other and clasped his hands round it, "they gave it me—for an unbirthday present."

"I beg your pardon?" Alice said with a puzzled air.
"I am not offended," said Humpty Dumpty.
"I mean, what *is* an unbirthday present?"
"A present given when it isn't your birthday, of course."
Alice considered a little. "I like birthday presents best," she said at last.
"You don't know what you're talking about!" cried Humpty Dumpty. "How many days are there in a year?"
"Three hundred and sixty-five," said Alice.
"And how many birthdays have you?"
"One."
"And if you take one from three hundred and sixty-five, what remains?"
"Three hundred and sixty-four, of course."
Humpty Dumpty looked doubtful. "I'd rather see that done on paper," he said.
Alice couldn't help smiling as she took out her memorandum book, and worked the sum for him:

$$\frac{\begin{array}{r}365\\1\end{array}}{364}$$

Humpty Dumpty took the book, and looked at it carefully. "That seems to be done right—" he began.
"You're holding it upside down!" Alice interrupted.
"To be sure I was!" Humpty Dumpty said gaily, as she turned it round for him. "I thought it looked a little queer. As I was saying, that *seems* to be done right—though I haven't time to look it over thoroughly just now—and that shows that there are three hundred and sixty-four days when you might get unbirthday presents—"
"Certainly," said Alice.
"And only one for birthday presents, you know. There's glory for you!"
"I don't know what you mean by 'glory,' " Alice said.
Humpty Dumpty smiled contemptuously. "Of course you don't—till I tell you. I meant 'there's a nice knockdown argument for you.' "
"But 'glory' doesn't mean 'a nice knockdown argument,' " Alice objected.
"When *I* use a word," Humpty Dumpty said in rather a scornful tone, "it means just what I choose it to mean—neither more nor less."
"The question is," said Alice, "whether you can make words mean so many different things."
"The question is," said Humpty Dumpty, "which is to be master—that's all."
Alice was too much puzzled to say anything, so, after a minute, Humpty Dumpty began again. "They've a temper, some of them—particularly verbs, they're the

proudest: adjectives you can do anything with, but not verbs. However, *I* can manage the whole lot of them! Impenetrability! That's what *I* say!"

"Would you tell me, please," said Alice, "what that means?"

"Now you talk like a reasonable child," said Humpty Dumpty, looking very much pleased, "I meant by 'impenetrability' that we've had enough of that subject, and it would be just as well if you'd mention what you mean to do next, as I suppose you don't mean to stop here all the rest of your life."

"That's a great deal to make one word mean," Alice said in a thoughtful tone.

"When I make a word do a lot of work like that," said Humpty Dumpty, "I always pay it extra."

"Oh!" said Alice. She was too much puzzled to make any other remark.

"Ah, you should see 'em come round me of a Saturday night," Humpty Dumpty went on, wagging his head gravely from side to side; "for to get their wages, you know."

(Alice did not venture to ask what he paid them with; and so, you see, I can't tell *you*.)

"You seem very clever at explaining words, sir," said Alice. "Would you kindly tell me the meaning of the poem called 'Jabberwocky'?"

"Let's hear it," said Humpty Dumpty. "I can explain all the poems that ever were invented—and a good many that haven't been invented just yet."

This sounded very hopeful, so Alice repeated the first verse:

> "'Twas brillig, and the slithy toves
> Did gyre and gimble in the wabe;
> All mimsy were the borogoves,
> And the mome raths outgrabe."

"That's enough to begin with," Humpty Dumpty interrupted; "there are plenty of hard words there. '*Brillig*' means four o'clock in the afternoon—the time when you begin *broiling* things for dinner."

"That'll do very well," said Alice. "And '*slithy*'?"

"Well, '*slithy*' means 'lithe and slimy.' 'Lithe' is the same as 'active.' You see, it's like a portmanteau—there are two meanings packed up in one word."

"I see it now," Alice remarked thoughtfully. "And what are '*toves*'?"

"Well, '*toves*' are something like badgers—they're something like lizards—and they're something like corkscrews."

"They must be very curious-looking creatures."

"They are that," said Humpty Dumpty, "also they make their nests under sundials—also they live on cheese."

"And what's to '*gyre*' and to '*gimble*'?"

"To '*gyre*' is to go round and round like a gyroscope. To '*gimble*' is to make holes like a gimlet."

"And '*the wabe*' is the grass-plot round a sundial, I suppose?" said Alice, surprised at her own ingenuity.

"Of course it is. It's called '*wabe*,' you know, because it goes a long way before it, and a long way behind it."

"And a long way beyond it on each side," Alice added.

"Exactly so. Well, then, '*mimsy*' is flimsy and miserable (there's another portmanteau for you). And a '*borogove*' is a thin, shabby-looking bird with its feathers sticking out all round—something like a live mop."

"And then '*mome raths*'?" said Alice. "I'm afraid I'm giving you a great deal of trouble."

"Well, a '*rath*' is a sort of green pig: but '*mome*' I'm not certain about. I think it's short for 'from home'—meaning that they'd lost their way, you know."

"And what does '*outgrabe*' mean?"

"Well, '*outgribing*' is something between bellowing and whistling, with a kind of sneeze in the middle; however, you'll hear it done, maybe—down in the wood yonder—and when you've once heard it you'll be *quite* content. Who's been repeating all that hard stuff to you?"

"I read it in a book," said Alice. "But I had some poetry repeated to me, much easier than that, by—Tweedledee, I think it was."

"As to poetry, you know," said Humpty Dumpty, stretching out one of his great hands, "*I* can repeat powetry as well as other folk, if it comes to that."

"Oh, it needn't come to that!" Alice hastily said, hoping to keep him from beginning.

"The piece I'm going to repeat," he went on, without noticing her remark, "was written entirely for your amusement."

Alice felt that in that case she really ought to listen to it, so she sat down, and said "Thank you" rather sadly.

> "In winter, when the fields are white,
> I sing this song for your delight—

only I don't sing it," he added, as an explanation.

"I see you don't," said Alice.

"If you can *see* whether I'm singing or not, you've sharper eyes than most," Humpty Dumpty remarked severely. Alice was silent.

"In spring, when woods are getting green,
I'll try and tell you what I mean."

"Thank you very much," said Alice.

"In summer, when the days are long,
Perhaps you'll understand the song.

In autumn, when the leaves are brown,
Take pen and ink, and write it down."

"I will, if I can remember it so long," said Alice.
"You needn't go on making remarks like that," Humpty Dumpty said; "they're not sensible, and they put me out.

"I sent a message to the fish:
I told them 'This is what I wish.'

The little fishes of the sea,
They sent an answer back to me.

The little fishes' answer was
'We cannot do it, Sir, because—'"

"I'm afraid I don't quite understand," said Alice.
"It gets easier further on," Humpty Dumpty replied.

"I sent to them again to say
'It will be better to obey.'

The fishes answered with a grin,
Why, what a temper you are in!'

I told them once, I told them twice;
They would not listen to advice.

I took a kettle large and new,
Fit for the deed I had to do.

My heart went hop, my heart went thump;
I filled the kettle at the pump.

Then someone came to me and said,
'The little fishes are in bed.'

I said to him, I said it plain,
'Then you must wake them up again.'

I said it very loud and clear;
I went and shouted in his ear."

Humpty Dumpty raised his voice almost to a scream as he repeated this verse, and Alice thought, with a shudder, "I wouldn't have been the messenger for *anything*!"

"But he was very stiff and proud;
He said, 'You needn't shout so loud!'

And he was very proud and stiff;
He said, 'I'd go and wake them, if—'

I took a corkscrew from the shelf;
I went to wake them up myself.

And when I found the door was locked,
I pulled and pushed and kicked and knocked.

And when I found the door was shut,
I tried to turn the handle, but—"

There was a long pause.
"Is that all?" Alice timidly asked.
"That's all," said Humpty Dumpty. "Good-by."

This was rather sudden, Alice thought; but, after such a *very* strong hint that she ought to be going, she felt that it would hardly be civil to stay. So she got up and held out her hand. "Good-by, till we meet again!" she said as cheerfully as she could.

"I shouldn't know you again if we *did* meet," Humpty Dumpty replied in a discontented tone, giving her one of his fingers to shake. "You're so exactly like other people."

"The face is what one goes by, generally," Alice remarked in a thoughtful tone.

"That's just what I complain of," said Humpty Dumpty. "Your face is the same as everybody has—the two eyes, so" (marking their places in the air with his thumb), "nose in the middle, mouth under. It's always the same. Now, if you had the two eyes on the same side of the nose, for instance—or the mouth at the top—that would be *some* help."

"It wouldn't look nice," Alice objected.

But Humpty Dumpty only shut his eyes and said, "Wait till you've tried."

Alice waited a minute to see if he would speak again, but as he never opened his eyes or took any further notice of her, she said "Good-by!" once more, and, getting no answer to this, she quietly walked away; but she couldn't help saying to herself as she went, "Of all the unsatisfactory" (she repeated this aloud, as it was a great comfort to have such a long word to say) "of all the unsatisfactory people I *ever* met—" She never finished the sentence, for at this moment a heavy crash shook the forest from end to end.

Chapter VIII

"It's My Own Invention"

... "The name of the song is called 'Haddocks' Eyes.'"

"Oh, that's the name of the song, is it?" Alice said, trying to feel interested.

"No, you don't understand," the Knight said, looking a little vexed. "That's what the name is *called*. The name really is '*The Aged, Aged Man.*'"

"Then I ought to have said, 'That's what the *song* is called'?" Alice corrected herself.

"No, you oughtn't; that's quite another thing! The *song* is called '*Ways and Means*'; but that's only what it's *called*, you know!"

"Well, what *is the* song, then?" said Alice, who was by this time completely bewildered.

"I was coming to that," the Knight said. "The song really *is* '*A-sitting on a Gate*'; and the tune's my own invention."

Questions on Carroll

1. Explain the problem of language being used as a comedy device by Carroll in Alice's conversation with the Red Queen.

2. Explain how Alice's encounter with Tweedledum and Tweedledee is an illustration of "logic" misused.

3. Can one believe impossible things? What is meant by impossible here?

4. Explain Humpty Dumpty's theory of definition. Compare it to Kant's.

5. After reading Humpty Dumpty's explanation of "Jabberwocky," find a word that is of the type Humpty explains and try to discover its hidden meaning. (Hint: start with chortle or frabjous or frumious.)

6. In the chapter named "It's My Own Invention," show two ways in which Alice confuses word with object.

7. Give some examples not found here of confusing word with object.

For some relevant extra reading for a possible term paper, see: W. V. O. Quine, "The Problem of Meaning in Linguistics." in *From a Logical Point of View*, Harper Torch Book, 1963, and L. Wittenstein, *The Blue and Brown Books*, Harper Torch Book, 1965.

Introduction

The next selection is excerpted from a classic in very early science fiction. Its author, Edwin A. Abbott, was a contemporary of Dodgson. Both of them lectured at major British Universities—Abbott at Cambridge and Dodgson at Oxford.

The styles of the two differ in important ways. While Carroll (Dodgson) was able to make creative use of non-sense, Abbott on the other hand sought to use fantasy to present his ideas, thereby creating a "sensible" fantasy. Being a product of his age (Victorian), Abbott could not resist moralizing, which in this instance does not hurt his work. His mathematical point is that the way we first view things may not be the way they really are (compare with Plato *et al.*). It was around this time that non-Euclidean geometry was beginning to cry out for recognition. Abbott here has provided us with a discussion of some of the joys and perils of considering novel ideas.

Flatland

A. Square

Part I

This World

Section 1—Of the Nature of Flatland. I call our world Flatland, not because we call it so, but to make its nature clearer to you, my happy readers, who are privileged to live in Space.

Imagine a vast sheet of paper on which straight Lines, Triangles, Squares, Pentagons, Hexagons, and other figures, instead of remaining fixed in their places, move freely about, on or in the surface, but without the power of rising above or sinking below it, very much like shadows—only hard and with luminous edges—and you will then have a pretty correct notion of my country and countrymen. Alas, a few years ago, I should have said "my universe": but now my mind has been opened to higher views of things.

In such a country, you will perceive at once that it is impossible that there should be anything of what you call a "solid" kind; but I dare say you will suppose that we could at least distinguish by sight the Triangles, Squares, and other figures, moving about as I have described them. On the contrary, we could see nothing of the kind, not at least so as to distinguish one figure from another. Nothing was visible, nor could be visible, to us, except Straight Lines; and the necessity of this I will speedily demonstrate.

Place a penny on the middle of one of your tables in Space; and leaning over it, look down upon it. It will appear a circle.

But now, drawing back to the edge of the table, gradually lower your eye (thus bringing yourself more and more into the condition of the inhabitants of Flatland), and you will find the penny becoming more and more oval to your view; and at last when you have placed your eye exactly on the edge of the table (so that you are, as it were, actually a Flatlander) the penny will then have ceased to appear oval at all, and will have become, so far as you can see, a straight line. . . .

. . . You may perhaps ask how under these disadvantageous circumstances we are able to distinguish our friends from one another; but the answer to this very natural question will be more fitly and easily given when I come to describe the inhabitants of Flatland. For the present let me defer this subject, and say a word or two about the climate and houses in our country.

Part 2

Of the Climate and Houses in Flatland

As with you, so also with us, there are four points of the compass: North, South, East and West.

There being no sun or other heavenly bodies, it is impossible for us to determine the North in the usual way; but we have a method of our own. By a Law of Nature with us, there is a constant attraction to the South; and, although in temperate climates this is very slight—so that even a Woman in reasonable health can journey several furlongs northward without much difficulty—yet the hampering effect of the southward attraction is quite sufficient to serve as a compass in most parts of our earth. Moreover, the rain (which falls at stated intervals) coming always from the North, is an additional assistance; and in the towns we have the guidance of the houses, which of course have their side-walls running for the most part North and South, so that the roofs may keep off the rain from the North. In the country, where there are no houses, the trunks of the trees serve as some sort of guide. Altogether, we have not so much difficulty as might be expected in determining our bearing.

Yet in our more temperate regions, in which the southward attraction is hardly felt, walking sometimes in a perfectly desolate plain where there have been no houses nor trees to guide me, I have been occasionally compelled to remain stationary for hours together, waiting till the rain came before continuing my journey. On the weak and aged, and especially delicate Females, the force of attraction tells much more heavily than on the robust of the Male Sex, so that it is a point of breeding, if you meet a Lady in the street, always to give her the North side of the way—by no means an easy thing to do always at short notice when you are in rude health and in a climate where it is difficult to tell your North from your South.

Windows there are none in our houses: for the light comes to us alike in our homes and out of them, by day and by night, equally at all times and in all places, whence we know not. It was in old days, with our learned men, an interesting and oft-investigated question, "What is the origin of light?" and the solution of it has been repeated attempted, with no other result than to crowd our lunatic asylums with the would-be solvers. Hence, after fruitless attempts to suppress such investigations indirectly by making them liable to a heavy tax, the Legislature, in comparatively recent times, absolutely prohibited them. I—alas, I alone in Flatland—know now only too well the true solution of this mysterious problem; but my knowledge cannot be made intelligible to a single one of my countrymen; and I am mocked at—I, the sole possessor of the truths of Space and of the theory of the introduction of Light from the world of three Dimensions—as if I were the maddest of the mad! But a truce to these painful digressions: let me return to our houses.

The most common form for the construction of a house is five-sided or pentagonal, as in the annexed figure. The two Northern sides RO, OF, constitute the roof, and for the most part have no doors; on the East is a small door for the Women; on the West a much larger one for the Men; the South side or floor is usually doorless.

Square and triangular houses are not allowed, and for this reason. The angles of a Square (and still more those of an equilateral Triangle), being much more pointed than those of a Pentagon, and the lines of inanimate objects (such as houses)

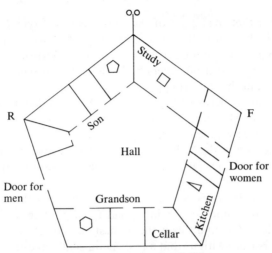

Figure 3-1. The House of a Square

being dimmer than the lines of Men and Women, it follows that there is no little danger lest the points of a square or triangular house residence might do serious injury to an inconsiderate or perhaps absentminded traveller suddenly running against them: and therefore, as early as the eleventh century of our era, triangular houses were universally forbidden by Law, the only exceptions being fortifications, powder-magazines, barracks, and other state buildings, which it is not desirable that the general public should approach without circumspection. . . .

Part 3

Concerning the Inhabitants of Flatland

The greatest length or breadth of a full grown inhabitant of Flatland may be estimated at about eleven of your inches. Twelve inches may be regarded as a maximum.

Our Women are Straight Lines.

Our Soldiers and Lowest Classes of Workmen are Triangles with two equal sides, each about eleven inches long, and a base of third side so short (often not exceeding half an inch) that they form at their vertices a very sharp and formidable angle. Indeed when their bases are of the most degraded type (not more than the eighth part of an inch in size), they can hardly be distinguished from Straight Lines of Women; so extremely pointed are their vertices. With us, as with you, these Triangles are distinguished from others by being called Isosceles; and by this name I shall refer to them in the following pages.

Our Middle Class consists of Equilateral or Equal-Sided Triangles.

Our Professional Men and Gentlemen are Squares (to which class I myself belong) and Five-Sided Figures or Pentagons.

Next above these come the Nobility, of whom there are several degrees, beginning at Six-Sided Figures, or Hexagons, and from thence rising in the number of their sides till they receive the honourable title of Polygonal, or many-sided. Finally when the number of the sides becomes so numerous, and the sides themselves so small, that the figure cannot be distinguished from a circle, he is included in the Circular or Priestly order; and this is the highest class of all.

It is a Law of Nature with us that a male child shall have one more side than his father, so that each generation shall rise (as a rule) one step in the scale of development and nobility. Thus the son of a Square is a Pentagon; the son of a Pentagon, a Hexagon; and so on.

But this rule applies not always to the Tradesman, and still less often to the Soldiers, and to the Workman; who indeed can hardly be said to deserve the name of human Figures, since they have not all their sides equal. With them therefore the Law of Nature does not hold; and the son of an Isosceles (i.e. a Triangle with two sides equal) remains Isosceles still. Nevertheless, all hope is not shut out, even from the Isosceles, that his posterity may ultimately rise above his degraded condition. For, after a long series of military successes, or diligent and skillful labours, it is generally found that the more intelligent among the Artisan and Soldier classes manifest a slight increase of their third side or base, and a shrinkage of the two other sides. Intermarriages (arranged by the Priests) between the sons and daughters of these more intellectual members of the lower classes generally result in an offspring approximating still more to the type of the Equal-Sided Triangle.

Rarely—in proportion to the vast numbers of Isosceles births—is a genuine and certifiable Equal-Sided Triangle produced from Isosceles parents.* Such a birth requires, as its antecedents, not only a series of carefully arranged intermarriages,

---

* "What need of a certificate?" a Spaceland critic may ask: "Is not the procreation of a Square Son a certificate from Nature herself, proving the Equal-sidedness of the Father?" I reply that no lady of any position will marry an uncertified Triangle. Square offspring has sometimes resulted from a slightly Irregular Triangle; but in almost every case the irregularity of the first generation is visited on the third; which either fails to attain the Pentagonal rank, or relapses to the Triangular.

but also a long-continued exercise of frugality and self-control on the part of the would-be ancestors of the coming Equilateral, and a patient, systematic, and continuous development of the Isosceles intellect through many generations . . . .

Part 4

Concerning the Women

If our highly pointed Triangles of the Soldier class are formidable, it may be readily inferred that far more formidable are our Women. For, if a Soldier is a wedge, a Woman is a needle, being, so to speak, *all* point, at least at the two extremities. Add to this the power of making herself practically invisible at will, and you will perceive that a Female, in Flatland, is a creature by no means to be trifled with . . . .

The dangers to which we are exposed from our Women must now be manifest to the meanest capacity in Spaceland. If even the angle of a respectable Triangle in the middle class is not without its dangers; if to run against a Working Man involves a gash; if collision with an Officer of the military class necessitates a serious wound; if a mere touch from the vertex of a Private Soldier brings with it danger of death;— what can it be to run against a Woman, except absolute and immediate destruction? And when a Woman is invisible, or visible only as a dim sub-lustrous point, how difficult must it be, even for the most cautious, always to avoid collision! . . . .

Part 5

Of Our Methods of Recognizing One Another

Recall what I told you above. All beings in Flatland, animate or inanimate, no matter what their form, present *to our view* the same, or nearly the same, appearance, viz. that of a straight line. How then can one be distinguished from another, where all appear the same?

The answer is threefold. The first means of recognition is the sense of hearing; which with us is far more highly developed than with you, and which enables us not only to distinguish by the voice our personal friends, but even to discriminate between different classes, at least so far as concerns the three lowest orders, the Equilateral, the Square, and the Pentagon—for of the Isosceles I take no account. But as we ascend in the social scale, the process of discriminating and being discriminated by hearing increases in difficulty, partly because voices are assimilated, partly because the faculty of voice-discrimination is a plebeian virtue not much developed among the Aristocracy. And wherever there is any danger of imposture we cannot trust to this method. Amongst our lowest orders, the vocal organs are developed to a degree more than correspondent with those of hearing, so that an Isosceles can easily feign the voice of a Polygon, and, with some training, that of a Circle himself. A second method is therefore more commonly resorted to.

Feeling is, among our Women and lower classes—about our upper classes I shall speak presently—the principal test of recognition, at all events between strangers, and when the question is, not as to the individual, but as to the class. What therefore "introduction" is among the higher classes in Spaceland, that the process of "feeling" is with us. "Permit me to ask you to feel and be felt by my friend Mr. So-and-so"—is still, among the more old-fashioned of our country gentlemen in districts remote from towns, the customary formula for a Flatland introduction. But in the towns, and among men of business, the words "be felt by" are omitted and the sentence is abbreviated to, "Let me ask you to feel Mr. So-and-so"; although it is assumed, of course, that the "feeling" is to be reciprocal. Among our still more modern and dashing young gentlemen—who are extremely averse to superfluous effort and supremely indifferent to the purity of their native language—the formula is still further curtailed by the use of "to feel" in a technical sense, meaning, "to recommend-for-the-purposes-of-feeling-and-being-felt"; and at this moment the "slang" of polite or fast society in the upper classes sanctions such a barbarism as "Mr. Smith, permit me to feel Mr. Jones." . . . .

Part 6

Of Recognition by Sight

. . . That this power exists in any regions and for any classes is the result of Fog. . . . If Fog were non-existent, all lines would appear equally and indistinguishably clear; and this is actually the case in those unhappy countries in which the atmosphere is perfectly dry and transparent. But wherever there is a rich supply of Fog, objects that are at a distance, say of three feet, are appreciably dimmer than those at a distance of two feet eleven inches; and the result is that by careful and constant experimental observation of comparative dimness and clearness, we are enabled to infer with great exactness the configuration of the object observed. . . .

Part 7

Concerning Irregular Figures

. . . Expediency therefore concurs with Nature in stamping the seal of its approval upon Regularity of conformation: nor has the Law been backward in seconding their efforts. "Irregularity of Figure" means with us the same as, or more than, a combination of moral obliquity and criminality with you, and is treated accordingly. There are not wanting, it is true, some promulgators of paradoxes who maintain that there is no necessary connection between geometrical and moral Irregularity. "The Irregular," they say, "is from his birth scouted by his own parents, derided by his brothers and sisters, neglected by the domestics, scorned and suspected by society, and excluded from all posts of responsibility, trust, and useful activity. His every movement is jealously watched by the police till he comes of age and presents himself for inspection; then he is either destroyed, if he is found to exceed the fixed margin of deviation, or else immured in a Government Office as a clerk of the seventh class; prevented from marriage; forced to drudge at an uninteresting occupation for a miserable stipend; obliged to live and board at the office, and to take even his vacation under close supervision; what wonder that human nature, even in the best and purest, is embittered and perverted by such surroundings!" . . . .

Part 15

Concerning a Stranger from Spaceland

. . . It was the last day of the 1999th year of our era. The pattering of the rain had long ago announced nightfall; and I was sitting in the company of my wife, musing on the events of the past and the prospects of the coming year, the coming century, the coming Millennium.

My four sons and two orphan Grandchildren had retired to their several apartments; and my wife alone remained with me to see the old Millennium out and the new one in.

I was rapt in thought, pondering in my mind some words that had casually issued from the mouth of my youngest Grandson, a most promising young Hexagon of unusual brilliancy and perfect angularity. His uncles and I had been giving him his usual practical lesson in Sight Recognition, turning ourselves upon our centres, now rapidly, now more slowly, and questioning him as to our positions; and his answers had been so satisfactory that I had been induced to reward him by giving him a few hints on Arithmetic, as applied to Geometry.

Taking nine Squares, each an inch every way, I had put them together so as to make one large Square, with a side of three inches, and I had hence proved to my little Grandson that—though it was impossible for us to *see* the inside of the Square—yet we might ascertain the number of square inches in a Square by simply squaring the number of inches in the side: "and thus," said I, "we know that 3^2, or 9, represents the number of square inches in a Square whose side is 3 inches long."

The little Hexagon meditated on this a while and then said to me; "But you have been teaching me to raise numbers to the third power: I suppose 3^3 must mean something in Geometry; what does it mean?" "Nothing at all," replied I, "not at least in Geometry; for Geometry has only Two Dimensions." And then I began to shew the boy how a Point by moving through a length of three inches makes a Line of three inches, which may be represented by 3; and how a Line of three inches, moving parallel to itself through a length of three inches, makes a Square of three inches every way, which may be represented by 3^2.

Upon this, my Grandson, again returning to his former suggestion, took me up rather suddenly and exclaimed, "Well, then, if a Point by moving three inches, makes a Line of three inches represented by 3; and if a straight Line of three inches, moving parallel to itself, makes a Square of three inches every way, represented by 3^2; it must be that a Square of three inches every way, moving somehow parallel to itself (but I don't see what) of three inches every way—and this must be represented by 3^3."

"Go to bed," said I, a little ruffled by this interruption: "if you would talk less nonsense, you would remember more sense."

So my Grandson had disappeared in disgrace; and there I sat by my Wife's side, endeavouring to form a retrospect of the year 1999 and of the possibilities of the year 2000, but not quite able to shake off the thoughts suggested by the prattle of my bright little Hexagon. Only a few sands now remained in the half-hour glass. Rousing myself from my reverie I turned the glass Northward for the last time in the old Millennium; and in the act, I exclaimed aloud, "The boy is a fool."

Straightway I became conscious of a Presence in the room, and a chilling breath thrilled through my very being. "He is no such thing," cried my Wife, "and you are breaking the Commandments in thus dishonouring your own Grandson." But I took no notice of her. Looking round in every direction I could see nothing; yet still I felt a Presence, and shivered as the cold whisper came again. I started up. "What is the matter?" said my Wife, "there is no draught; what are you looking for? There is nothing." There was nothing; and I resumed my seat, again exclaiming, "The boy is a fool, I say; 3^3 can have no meaning in Geometry." At once there came a distinctly audible reply, "The boy is not a fool; and 3^3 has an obvious Geometrical meaning."

My wife as well as myself heard the words, although she did not understand their meaning, and both of us sprang forward in the direction of the sound. What was our horror when we saw before us a Figure! At the first glance it appeared to be a Woman, seen sideways; but a moment's observation shewed me that the extremities passed into dimness too rapidly to represent one of the Female Sex; and I should have thought it a Circle, only that it seemed to change its size in a manner impossible for a Circle or for any regular Figure of which I had had experience.

But my Wife had not my experience, nor the coolness necessary to note these characteristics. With the usual hastiness and unreasoning jealousy of her Sex, she flew at once to the conclusion that a Woman had entered the house through some small aperture. "How comes this person here?" she exclaimed, "you promised me, my dear, that there should be no ventilators in our new house." "Nor are there any," said I; "but what makes you think that the stranger is a Woman? I see by my power of Sight Recognition—" "Oh, I have no patience with your Sight Recognition," replied she, " 'Feeling is believing' and 'A Straight Line to the touch is worth a Circle to the sight' "—two Proverbs, very common with the Frailer Sex in Flatland.

"Well," said I, for I was afraid of irritating her, "if it must be so, demand an introduction." Assuming her most gracious manner, my Wife advanced towards the Stranger, "Permit me, Madam, to feel and be felt by—" then, suddenly recoiling, "Oh! it is not a Woman, and there are no angles either, not a trace of one. Can it be that I have so misbehaved to a perfect Circle?"

"I am indeed, in a certain sense a Circle," replied the Voice, "and a more perfect Circle than any in Flatland; but to speak more accurately, I am many Circles in one." Then he added more mildly, "I have a message, dear Madam, to your husband, which I must not deliver in your presence; and, if you would suffer us to retire for a few minutes—" But my Wife would not listen to the proposal that our august Visitor should so incommode himself, and assuring the Circle that the hour of her own retirement had long passed, with many reiterated apologies for her recent indiscretion, she at last retreated to her apartment.

I glanced at the half-hour glass. The last sands had fallen. The third Millennium had begun.

Part 16

How the Stranger Vainly Endeavoured to Reveal to Me in Words the Mysteries of Spaceland

... *I.* Most illustrious Sir, excuse my awkwardness, which arises not from ignorance of the usages of polite society, but from a little surprise and nervousness, consequent on this somewhat unexpected visit. And I beseech you to reveal my indiscretion to no one, and especially not to my Wife. But before your Lordship enters into further communications, would he deign to satisfy the curiosity of one who would gladly know whence his visitor came?

Stranger. From Space, from Space, Sir: whence else?

I. Pardon me, my Lord, but is not your Lordship already in Space, your Lordship and his humble servant, even at this moment?

Stranger. Pooh! what do you know of Space? Define Space.

I. Space, my Lord, is height and breadth indefinitely prolonged.

Stranger. Exactly: you see you do not even know what Space is. You think it is of Two Dimensions only; but I have come to announce to you a Third—height, breadth, and length....

I. Would your Lordship indicate or explain to me in what direction is the Third Dimension, unknown to me?

Stranger. I came from it. It is up above and down below.

I. My Lord means seemingly that it is Northward and Southward.

Stranger. I mean nothing of the kind. I mean a direction in which you cannot look, because you have no eye in your side.

I. Pardon me, my Lord, a moment's inspection will convince your Lordship that I have a perfect luminary at the juncture of two of my sides.

Stranger. Yes: but in order to see into Space you ought to have an eye, not on your Perimeter, but on your side, that is, on what you would probably call your inside; but we in Spaceland should call it your side.

I. An eye in my inside! An eye in my stomach! Your Lordship jests.

Stranger. I am in no jesting humour. I tell you that I come from Space, or, since you will not understand what Space means, from the Land of Three Dimensions whence I but lately looked down upon your Plane which you call Space forsooth.

From that position of advantage I discerned all that you speak of as *solid* (by which you mean "enclosed on four sides"), your houses, your churches, your very chests and safes, yes even your insides and stomachs, all lying open and exposed to my view.

I.   Such assertions are easily made, my Lord.

Stranger.   But not easily proved, you mean. But I mean to prove mine . . . . You are living on a Plane. What you style Flatland is the vast level surface of what I may call a fluid, on, or in, the top of which you and your countrymen move about, without rising above it or falling below it.

I am not a plane Figure, but a Solid. You call me a Circle; but in reality I am not a Circle, but an infinite number of Circles, of size varying from a Point to a Circle of thirteen inches in diameter, one placed on the top of the other. When I cut through your plane as I am now doing, I make in your plane a section which you, very rightly, call a Circle. For even a Sphere—which is my proper name in my own country—if he manifest himself at all to an inhabitant of Flatland—must needs manifest himself as a Circle . . . .

The diminished brightness of your eyes indicates incredulity. But now prepare to receive proof positive of the truth of my assertions. You cannot indeed see more than one of my sections, or Circles, at a time; for you have no power to raise your eye out of the plane of Flatland; but you can at least see that, as I rise in Space, so my sections become smaller. See now, I will rise; and the effect upon your eye will be that my Circle will become smaller and smaller until it dwindles to a point and finally vanishes.

There was a "rising" that I could see; but he diminished and finally vanished. I winked once or twice to make sure that I was not dreaming. But it was no dream. For from the depths of nowhere came forth a hollow voice—close to my heart it seemed— "Am I quite gone? Are you convinced now? Well, now I will gradually return to Flatland and you shall see my section become larger and larger." . . .

"Monster," I shrieked, "be thou juggler, enchanter, dream, or devil, no more will I endure thy mockeries. Either thou or I must perish." And saying these words I precipitated myself upon him.

Part 17

How the Sphere, Having in Vain Tried Words, Resorted to Deeds

. . . *Sphere.*   Why will you refuse to listen to reason? I had hoped to find in you— as being a man of sense and an accomplished mathematician—a fit apostle for the

Gospel of the Three Dimensions, which I am allowed to preach once only in a thousand years: but now I know not how to convince you. Stay, I have it. Deeds, and not words, shall proclaim the truth. Listen, my friend,"... thundered the Stranger: "then meet your fate: out of your Plane you go. Once, twice, thrice! 'Tis done!"

Part 18

How I Came to Spaceland, and What I Saw There

An unspeakable horror seized me. There was a darkness; then a dizzy, sickening sensation of sight that was not like seeing; I saw a Line that was no Line; Space that was not Space: I was myself, and not myself. When I could find voice, I shrieked aloud in agony, "Either this is madness or it is Hell." "It is neither," calmly replied the voice of the Sphere, "it is Knowledge; it is Three Dimensions: open your eye once again and try to look steadily."

I looked, and, behold, a new world! There stood before me, visibly incorporate, all that I had before inferred, conjectured, dreamed, of perfect Circular beauty....

The Sphere would willingly have continued his lessons by indoctrinating me in the conformation of all regular Solids, Cylinders, Cones, Pyramids, Pentahedrons, Hexahedrons, Dodecahedrons, and Spheres: but I ventured to interrupt him. Not that I was wearied of knowledge. On the contrary, I thirsted for yet deeper and fuller draughts than he was offering to me.

"Pardon me," said I, "O Thou Whom I must no longer address as the Perfection of all Beauty; but let me beg thee to vouchsafe thy servant a sight of thine interior."

Sphere. My what?

I. Thine interior: thy stomach, thy intestines.

Sphere. Whence this ill-timed impertinent request? And what mean you by saying that I am no longer the Perfection of all Beauty?

I. My Lord, your own wisdom has taught me to aspire to One even more great, more beautiful, and more closely approximate to Perfection than yourself. As you yourself, superior to all Flatland forms, combine many Circles in One, so doubtless there is One above you who combines many Spheres in One Supreme Existence, surpassing even the Solids of Spaceland. And even as we, who are now in Space, look down on Flatland and see the insides of all things, so of a certainty there is yet above us some higher, purer region, whither thou dost surely purpose to lead me—O Thou Whom I shall always call, everywhere and in all Dimensions, my Priest, Philosopher, and Friend—some yet more spacious space, some more dimensionable Dimensionality, from the vantage-ground of which we shall look down together upon the revealed insides of Solid things, and where thine own intestines, and those of thy kindred Spheres, will lie exposed to the view of the poor wandering exile from Flatland, to whom so much has already been vouchsafed.

Sphere. Pooh! Stuff! Enough of this trifling! The time is short, and much remains to be done before you are fit to proclaim the Gospel of Three Dimensions to your blind benighted countrymen in Flatland.

I. Nay, gracious Teacher, deny me not what I know it is in thy power to perform. Grant me but one glimpse of thine interior, and I am satisfied for ever, remaining henceforth thy docile pupil, thy unemancipable slave, ready to receive all thy teachings and to feed upon the words that fall from thy lips.

Sphere. Well, then, to content and silence you, let me say at once, I would shew you what you wish if I could; but I cannot. Would you have me turn my stomach inside out to oblige you?

I. But my Lord has shewn me the intestines of all my countrymen in the Land of Two Dimensions by taking me with him into the Land of Three. What therefore more easy than now to take his servant on a second journey into the blessed region of the Fourth Dimension, where I shall look down with him once more upon this land of Three Dimensions, and see the inside of every three-dimensioned house, the secrets of the solid earth, the treasures of the mines in Spaceland, and the intestines of every solid living creature, even of the noble and adorable Spheres.

Sphere. But where is this land of Four Dimensions?

I. I know not: but doubtless my Teacher knows.

Sphere. Not I. There is no such land. The very idea of it is utterly inconceivable.

I. Not inconceivable, my Lord, to me, and therefore still less inconceivable to my Master. Nay, I despair not that, even here, in this region of Three Dimensions, your Lordship's art may make the Fourth Dimension visible to me; just as in the Land of Two Dimensions my Teacher's skill would fain have opened the eyes of his blind servant to the invisible presence of a Third Dimension, though I saw it not . . . .

I. If it indeed be so, that this other Space is really Thoughtland, then take me to that blessed Region where I in Thought shall see the insides of all solid things. There, before my ravished eye, a Cube moving in some altogether new direction, but strictly according to Analogy, so as to make every particle of his interior pass through a new kind of Space, with a wake of its own—shall create a still more perfect perfection than himself, with sixteen terminal Extra-solid angles, and Eight solid Cubes for his Perimeter. And once there, shall we stay our upward course? In that blessed region of Four Dimensions, shall we linger on the threshold of the Fifth, and not enter therein? Ah, no! Let us rather resolve that our ambition shall soar with our corporal ascent.

Then, yielding to our intellectual onset, the gates of the Sixth Dimension shall fly open; after that a Seventh, and then an Eighth—

How long I should have continued I know not. In vain did the Sphere, in his voice of thunder, reiterate his command of silence, and threaten me with the direst penalties if I persisted. Nothing could stem the flood of my ecstatic aspirations. Perhaps I was to blame; but indeed I was intoxicated with the recent draughts of Truth to which he himself had introduced me. However, the end was not long in coming. My words were cut short by a crash outside, and a simultaneous crash inside me, which impelled me through space with a velocity that precluded speech. Down! down! down! I was rapidly descending; and I knew that return to Flatland was my doom. One glimpse, one last and never-to-be-forgotten glimpse, I had of that dull level wilderness—which was now to become my Universe again—spread out before my eye. Then a darkness. Then a final, all-consummating thunder-peal; and, when I came to myself, I was once more a common creeping Square, in my Study at home, listening to the Peace-Cry of my approaching Wife . . . .

Part 21
How I Tried to Teach the Theory of Three Dimensions to My Grandson and with What Success

I awoke rejoicing, and began to reflect on the glorious career before me. I would go forth, methought, at once, and evangelize the whole of Flatland. Even to Women and Soldiers should the Gospel of Three Dimensions be proclaimed. I would begin with my Wife.

Just as I had decided on the plan of my operations, I heard the sound of many voices in the street commanding silence. Then followed a louder voice. It was a herald's proclamation. Listening attentively, I recognized the words of the Resolution of the Council, enjoining the arrest, imprisonment, or execution of any one who should pervert the minds of the people by delusions, and by professing to have received revelations from another World . . . .

My Pentagonal Sons were men of character and standing, and physicians of no mean reputation, but not great in mathematics, and, in that respect, unfit for my purpose. But it occurred to me that a young and docile Hexagon, with a mathematical turn, would be a most suitable pupil. Why therefore not make my first experiment with my little precocious Grandson, whose casual remarks on the meaning of 3^3 had met with the approval of the Sphere? Discussing the matter with him, a mere boy, I should be in perfect safety; for he would know nothing of the Proclamation of the

Council; whereas I could not feel sure that my Sons—so greatly did their patriotism and reverence for the Circles predominate over mere blind affection—might not feel compelled to hand me over to the Prefect, if they found me seriously maintaining the seditious heresy of the Third Dimension . . . .

When my Grandson entered the room I carefully secured the door. Then, sitting down by his side and taking our mathematical tablets, —or, as you would call them, Lines—I told him we would resume the lesson of yesterday. I taught him once more how a Point by motion in One Dimension produces a Line, and how a straight Line in Two Dimensions produces a Square. After this, forcing a laugh, I said, "And now, you scamp, you wanted to make me believe that a Square may in the same way by motion 'upward, not Northward' produce another figure, a sort of extra Square in Three Dimensions. Say that again, you young rascal."

At this moment we heard once more the herald's "O yes! O yes!" outside in the street proclaiming the Resolution of the Council. Young though he was, my Grandson—who was unusually intelligent for his age, and bred up in perfect reverence for the authority of the Circles—took in the situation with an acuteness for which I was quite unprepared. He remained silent till the last words of the Proclamation had died away, and then, bursting into tears, "Dear Grandpapa," he said, "that was only my fun, and of course I meant nothing at all by it; and we did not know anything about the Third Dimension; and I am sure I did not say one word about 'Upward, not Northward,' for that would be such nonsense, you know. How could a thing move Upward, and not Northward? Upward and not Northward! Even if I were a baby, I could not be so absurd as that. How silly it is! Ha! ha! ha!"

"Not at all silly," said I, losing my temper; "here for example, I take this Square," and, at the word, I grasped a moveable Square, which was lying at hand— "and I move it, you see, not Northward but—yes, I move it Upward—that is to say, Northward but I move it somewhere—not exactly like this, but somehow— " Here I brought my sentence to an inane conclusion, shaking the Square about in a purposeless manner, much to the amusement of my Grandson, who burst out laughing louder than ever, and declared that I was not teaching him, but joking with him; and so saying he unlocked the door and ran out of the room. Thus ended my first attempt to convert a pupil to the Gospel of Three Dimensions.

Part 22

How I Tried to Diffuse the Theory of Three Dimensions by Other Means, and of the Result

My failure with my Grandson did not encourage me to communicate my secret to others of my household; yet neither was I led by it to despair of success. Only I saw that I must not wholly rely on the catch-phrase, "Upward, not Northward," but must rather endeavour to seek a demonstration by setting before the public a clear view of the whole subject; and for this purpose it seemed necessary to resort to writing.

So I devoted several months in privacy to the composition of a treatise on the mysteries of Three Dimensions . . . .

One day, about eleven months after my return from Spaceland, I tried to see a Cube with my eye closed, but failed; and though I succeeded afterwards, I was not then quite certain (nor have I been ever afterwards) that I had exactly realized the original. This made me more melancholy than before, and determined me to take some step; yet what, I knew not. I felt that I would have been willing to sacrifice my life for the Cause, if thereby I could have produced conviction. But if I could not convince my Grandson, how could I convince the highest and most developed Circles in the land?

And yet at times my spirit was too strong for me, and I gave vent to dangerous utterances. Already I was considered heterodox if not treasonable, and I was keenly alive to the danger of my position; nevertheless I could not at times refrain from bursting out into suspicious or half-seditious utterances, even among the highest Polygonal and Circular society. When, for example, the question arose about the treatment of those lunatics who said that they had received the power of seeing the insides of things, I would quote the saying of an ancient Circle, who declared that prophets and inspired people are always considered by the majority to be mad; and I could not help occasionally dropping such expressions as "the all-seeing land"; once or twice I even let fall the forbidden terms "the Third and Fourth Dimensions."

At last, to complete a series of minor indiscretions, at a meeting of our Local Speculative Society held at the palace of the Prefect himself, —some extremely silly person having read an elaborate paper exhibiting the precise reasons why Providence has limited the number of Dimensions to Two, and why the attribute of omnividence is assigned to the Supreme alone—I so far forgot myself as to give an exact account of the whole of my voyage with the Sphere into Space, and to the Assembly Hall in our Metropolis, and then to Space again, and of my return home, and of everything that I had seen and heard in fact or vision. At first, indeed, I pretended that I was describing the imaginary experiences of a fictitious person; but my enthusiasm soon forced me to throw off all disguise, and finally, in a fervent peroration, I exhorted all my hearers to divest themselves of prejudice and to become believers in the Third Dimension.

Need I say that I was at once arrested and taken before the Council?

Next morning, standing in the very place where but a very few months ago the Sphere had stood in my company, I was allowed to begin and to continue my narration unquestioned and uninterrupted. But from the first I foresaw my fate; for the President, noting that a guard of the better sort of Policemen was in attendance, of angularity little, if at all, under 55, ordered them to be relieved before I began my defence, by an inferior class of 2 or 3. I knew only too well what that meant. I was to be executed or imprisoned, and my story was to be kept secret from the world by the simultaneous destruction of the officials who had heard it; and, this being the case, the President desired to substitute the cheaper for the more expensive victims.

After I had concluded my defence, the President, perhaps perceiving that some of the junior Circles had been moved by my evident earnestness, asked me two questions:—

1. Whether I could indicate the direction which I meant when I used the words "Upward, not Northward"?
2. Whether I could by any diagrams or descriptions (other than the enumeration of imaginary sides and angles. indicate the Figure I was pleased to call a Cube?

I declared that I could say nothing more, and that I must commit myself to the Truth, whose cause would surely prevail in the end.

The President replied that he quite concurred in my sentiment, and that I could not do better. I must be sentenced to perpetual imprisonment; but if the Truth intended that I should emerge from prison and evangelize the world, the Truth might be trusted to bring that result to pass. Meanwhile I should be subjected to no discomfort that was not necessary to preclude escape, and, unless I forfeited the privilege by misconduct, I should be occasionally permitted to see my brother who had preceded me to my prison.

Seven years have elapsed and I am still a prisoner, and—if I except the occasional visits of my brother—debarred from all companionship save that of my jailers. My brother is one of the best of Squares, just, sensible, cheerful, and not without fraternal affection; yet I confess that my weekly interviews, at least in one respect, cause me the bitterest pain. He was present when the Sphere manifested himself in the Council Chamber; he saw the Sphere's changing sections; he heard the explanation of the phenomena then given to the Circles. Since that time, scarcely a week has passed during seven whole years, without his hearing from me a repetition of the part I played in that manifestation, together with ample descriptions of all the phenomena in Spaceland, and the arguments for the existence of Solid things derivable from Analogy. Yet—I take shame to be forced to confess it—my brother has not yet grasped the nature of the Third Dimension, and frankly avows his disbelief in the existence of a Sphere.

Hence I am absolutely destitute of converts, and, for aught that I can see, the millennial Revelation has been made to me for nothing. Prometheus up in Spaceland was bound for bringing down fire for mortals, but I—poor Flatland Prometheus—lie here in prison for bringing down nothing to my countrymen. Yet I exist in the hope that these memoirs, in some manner, I know not how, may find their way to the minds of humanity in Some Dimension, and may stir up a race of rebels who shall refuse to be confined to limited Dimensionality.

That is the hope of my brighter moments. Alas, it is not always so. Heavily weighs on me at times the burdensome reflection that I cannot honestly say I am confident as to the exact shape of the once-seen, oft-regretted Cube; and in my nightly visions the mysterious precept, "Upward, not Northward," haunts me like a soul-devouring Sphinx. It is part of the martyrdom which I endure for the cause of the Truth that there are seasons of mental weakness, when Cubes and Spheres flit away into the background of scarce-possible existences; when the Land of Three Dimensions seems almost as visionary as the Land of One or None; nay, when even this hard wall that bars me from my freedom, these very tablets on which I am writing, and all the substantial realities of Flatland itself, appear no better than the offspring of a diseased imagination, or the baseless fabric of a dream.

Questions on Flatland

1. What are some of the sociological problems raised?

2. In Flatland, how is one's stratum determined? In what sense is the determination mathematically based? What are the parallels with our social structure?

3. Call the criteria for social position axioms; from where do the axioms come? That is, are they inductively or deductively originated?

4. How are a square's difficulties in visualizing three dimensions parallel to those of a three-dimensional dweller trying to visualize four-space?

5. Those of you who are science fiction fans or have had some "Theory of Relativity" experience should note that this book was written around fifty years before Einstein. Compare the sphere's apparent change in size with the similar phenomena in Einstein's theory.

6. Compare *Flatland* with "The Allegory of the Cave"; compare each with Alice's encounter with the Red Queen.

Part Four

What If _____ ?

Introduction

These next two articles allow you, the reader, to participate. Mathematics, to be fully appreciated, must be actively participated in. For mathematics to become real to you, you must do some.

These articles serve another purpose. At the heart of every mathematical system is a set of axioms. Consequently, mathematics is said to use the axiomatic method. However, whence come these axioms? How does one know when one has a good set of axioms? What is the relationship of a set of axioms to some "real" situation? Unfortunately, no answers to these questions seem completely satisfactory. However, some partial and somewhat practical answers are given in the following articles. The phrase "axiom system" can refer to either of two types of systems (though some of the same rules apply to both). On the one hand, are systems that seem to appear with little or no physical motivation; once they have come into being, someone may try to fit concrete situations to them. These are often called *formal systems* because of the emphasis on form rather than content. The first article by Montague and Montgomery discusses such systems.

A second point of view is expressed by Benbow and Goedicke in the second article. Here regularity is observed in a "real" situation, and an attempt is made to express these observations in precise form. Since all possible cases can be covered, the inductive process can be called *complete*. The axioms then have content from their inception. Axiom systems of this type are often called *material systems*.

I use a slightly different approach in the text that accompanies this reading book. That approach leads to a more natural terminology than the somewhat classical terms just introduced. In the text, axiom systems are classified as syntactic and semantic. I leave to the interested reader upon completion of his reading the question of why these terms were chosen.

How Mathematicians Develop a Branch of Pure Mathematics

Harriet F. Montague and Mabel D. Montgomery

Introduction to the Axiomatic Method

In this chapter we study the *axiomatic method.* Those who have studied geometry have seen this method used by the authors of their textbooks. The first planned use of this method in mathematics dates back to the time of Euclid (300 B.C.). Since the emphasis in mathematics at that time was on geometry, it is not surprising that the emphasis on the axiomatic method in the schools has traditionally been in the field of geometry. Today's students are being introduced to the axiomatic method in algebra as well as in geometry.

The axiomatic method is the framework within which all the "If–then" statements of mathematics are formed. The process of arriving at such a statement involves not only the hypothesis and the conclusion of the "If–then" statement itself; but also the laws of logic and the use of previously proved theorems, definitions, and, ultimately, certain basic statements called variously *axioms, postulates, assumptions.* To have a starting place, it is necessary to have such assumptions (axioms, postulates) and to accept them without attempts at proof. If we insisted on a proof for every statement we made, we would be involved in an unending regression. This framework developed from a set of basic assumptions is not peculiar to mathematics. Our everyday conduct is based on certain assumptions. Each of us behaves in a certain way because each of us abides by convictions or self-determined rules. A student does not cheat if he has a conviction against dishonesty. A person pays his bills because he has agreed to certain laws which compel him to pay or to suffer punishment, and he does

From *The Significance of Mathematics* by H. F. Montague and M. D. Montgomery, pp. 119–126, and 133–135. Charles E. Merrill Books, Inc., 1963. Reprinted by permission.

not wish to be punished. A community makes certain laws which form the basis for the conduct of community affairs.

Consider the statement "If I do not pay my federal income tax, then I am liable to fine and/or imprisonment." Why do we make such statements? Because our national economy is based on certain laws agreed on by authorized agencies, and penalties for noncompliance have also been agreed on. In addition, there are factors involving our obligations as citizens of the United States. Behind the statement, then, there are basic assumptions acting as "rules of the game." Consider another statement. "If I go out in the rain without proper protection, then I am liable to catch cold." This statement is based on other statements such as "Rain is wet," "Being wet causes changes in body temperature," "Certain conditions of rapid changes in body temperature, combined with other factors, provide a favorable climate for virus infections," etc.

In these nonmathematical situations we have:

(1) the presence of basic assumptions,
(2) the acceptance of these basic assumptions.

An individual's code of ethics is an example of a set of basic assumptions. It forms for him a set of axioms from which he derives the theorems determining his conduct. The Bill of Rights in our Constitution can be thought of as a set of postulates on which the laws of our country are based. An understanding of this concept of basic assumptions would do much to eliminate futile arguments between individuals and between groups of persons. Did you ever argue with another person and not come to a common meeting place? Did you ever stop to analyze why you never agreed? What were your basic assumptions? What were his? If they were contradictory sets of assumptions, then you should not have expected agreement. Even though each of you used valid reasoning in your arguments, you should have expected disagreement. Until equivalent basic assumptions are made by both parties, agreement cannot be expected. This concept applies to world affairs as well as to the affairs of individuals. Hence, we see that the conflict of political ideologies can be expected to continue as long as basic assumptions are contradictory.

In mathematics, the "cold war" of conflicting sets of assumptions has proved fruitful and exciting. Mathematicians realize that the conflicting sets of assumptions are incompatible, but they are willing to let the proponents of the conflicting sets develop the theorems which follow from each set of assumptions. Mathematician A may start with his set of assumptions, or axioms, and mathematician B may start with his set of axioms. A's axioms lead to some theorems. B's axioms lead to theorems. A does not quarrel with B's theorems unless B's reasoning is incorrect. B is tolerant of A's theorems as long as A has used valid reasoning. A and B both realize that if their sets of axioms are not in agreement, then they must expect conflicting theorems. *Neither A nor B would deny the propriety or possible usefulness of the other's theorems.* This freedom now enjoyed by mathematicians to develop theorems from many sets of axioms is a recent development in mathematics. We shall have more to say about it in a later chapter.

Consistency of a Set of Axioms

Our starting place in developing a branch of pure mathematics is a set of axioms. (Alternate words for axioms are *postulates, assumptions*. We shall use the word axioms in this work.) This set of axioms forms the set of rules of the game, and the rules are to be accepted without question. If the rules are changed, a different game results. In like manner, if the set of axioms is changed, a different branch of mathematics is developed.

The set of axioms must be chosen so that it contains no contradictions. This insures that no contradictory theorems will be derived from the set of axioms. We would not want, for example, to have one axiom saying "A circle is round" and another axiom saying "A circle is square" in the same set of axioms. A set of axioms which contains no contradictions is said to be *consistent*. Consistency is a required property of any set of axioms.

The axioms are statements and statements involve words. Just as it is impossible to prove every statement, so is it impossible to define every word. As a result, some words must remain undefined. This is necessary in order to avoid an infinite regression of definitions. These undefined words in the axioms form the undefined concepts in the branch of mathematics being developed. Such concepts in geometry are "point" and "line." They remain undefined. Other words in the axioms are words of ordinary language such as "and" and "exactly" and they are given the same meanings as they have in ordinary language.

Perhaps it seems strange to you for us to say that words like "point" and "line" are undefined, since most people think of a point as a dot made by a pencil or a piece of chalk, and a line as something drawn by a pencil against the side of a ruler. This reflects our desire to interpret geometric concepts in terms of the world around us. The first people who developed geometry had this same desire and for centuries no one thought of point and line in any other way. But together with the freedom to choose various sets of axioms came the freedom from being bound to the world of objects we see around us. Points and lines can have many interpretations, as we shall see in working with various sets of axioms.

The problem of showing that a given set of axioms is consistent is met by the use of *models*. A model for consistency is a concrete, physical interpretation of the undefined concepts, constructed in such a way that all the axioms of a given set of axioms are satisfied. By *satisfied*, we mean that the truth value of each one of the axioms is T. If we can construct such a model, the axioms are said to be *consistent*. In most cases, many models can be constructed if one can be. One of the appealing features of testing for consistency is this very large potential supply of models. Anyone with a creative urge usually enjoys the construction of these models.

The first set of axioms we shall examine for consistency is a set of three statements about the undefined concepts *ogg* and *uff*. We use these nonsense words purposely so that no one will have any preconceived notions about them. Also it must be remembered that we are accepting these statements without proof, if they are to serve as axioms. Let us call this set of axioms the axiom set A. The axiom set A is as follows:

Axiom 1. There are exactly three oggs.
Axiom 2. There is at least one uff.
Axiom 3. There are exactly two oggs on each uff.

It is our task to test the consistency of this set of axioms. The undefined concepts ogg and uff can be interpreted in any way we wish, but we will try to interpret ogg and uff in such a way that we can construct a model of oggs and uffs satisfying the three axioms.

Interpret an ogg to be a house and represent it in our model by a cross (X). Interpret an uff to be a street and represent it in our model by a line segment (_____). Figure 4-1 shows a model with these interpretations of ogg and uff.

Is axiom 1 satisfied by this model? Yes, there are exactly three crosses, or houses—no more, no less.

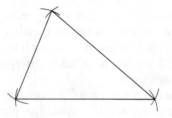

Figure 4-1

Is axiom 2 satisfied? Yes, there is at least one street. The fact that there are three streets in the model does not make any difference in testing this axiom—all we need is one street in order to have the axiom satisfied.

Is axiom 3 satisfied? Yes, there are exactly two houses on each street.

We have, then, a consistent set of axioms in axiom set A.

Consider next axiom set B, in which "point" and "line" are the undefined terms:

Axiom 1. There are exactly three points.
Axiom 2. There is at least one line.
Axiom 3. There are exactly two points on each line.

Instead of interpreting point as a dot and lines as something drawn by using a ruler, we shall interpret a point as a capital letter such as M, X, or T, and line as a succession of capital letters such as MRT, $XYZW$, etc. If a capital letter appears in such a succession, the point designated by the letter shall be interpreted to be on the line represented by the succession of letters. For example, the point R is on the line MRT, as are the points M and T. Now for our model to show the consistency of axiom set B, we display the lines PM, MO, and OP. In this model there are exactly three points P, M, and O, since there are just these three capital letters used in the model. Also there is at least one line (there are, in fact, three lines). Finally, there are exactly two points on each line: on line PM there are the two points P and M, on line MO there are the two points M and O, on line OP there are the two points O and P.

The model just used to show that axiom set B is consistent demonstrates that not all models need to be geometric. Moreover, the undefined concepts can be interpreted in any way our fancy dictates. What we must remember is that our interpretations of the undefined concepts are supposed to help us to build models which satisfy the axioms. Thus, some measure of restraint must be exercised.

Perhaps you have observed a relation between axiom set A and axiom set B. If the word "ogg" is replaced by "point" and the word "uff" by "line," axiom set A becomes axiom set B.

Now consider axiom set C, in which "bird" and "nest" are undefined:

Axiom 1. Every bird lives in a nest.

Axiom 2. Each bird lives in a nest with exactly one other bird.

Axiom 3. Not all birds live in the same nest.

Axiom 4. No bird lives in more than one nest.

In building a consistency model for axiom set C, we interpret a bird to be a dot (.) and a nest to be what we ordinarily think of as a line. Let us examine a model made up of two parallel lines, each with two dots on it (Figure 4-2).

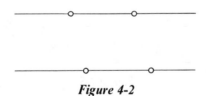

Figure 4-2

Axiom 1 of axiom set C is satisfied, since each dot is on a line. Axiom 2 is satisfied, since each dot is on a line with just one other dot. Not all dots are on the same line, so axiom 3 is satisfied. No dot is on more than one line in this model, so axiom 4 is satisfied. Axiom C is consistent, since the model satisfies all the axioms, with the interpretations of bird and nest as given.

A second model for axiom set C can be constructed by interpreting bird as a capital letter and nest as a sequence of capital letters in much the same way as we did for axiom set B. This new model is made up of the nests PT and RS. It can be verified that this simple model satisfies all the axioms of axiom set C.

Of course we wish to do more with axiom sets than to build models to show consistency. What the mathematician does after he forms a consistent set of axioms is to try to discover theorems; i.e., other statements which can be derived from that set of axioms. If we used axiom set A, we would want some theorems about oggs and uffs. Axiom set C should give us some theorems about birds and nests. Sometimes hunches about possible theorems are obtained from consistency models. We might have a hunch about possible theorems about birds and nests from our consistency model for axiom set C. We might be tempted to state:

Potential Theorem 1. There are exactly four birds.
Potential Theorem 2. There are exactly two nests.

This is a legitimate way to get ideas for theorems. Unfortunately, theorems are usually harder to come by. Another consistency model for axiom set C will show that potential theorems 1 and 2 can never be theorems. Figure 4-3 shows this other consistency model for axiom set C.

Figure 4-3

The property of consistency, as we have pointed out, is a "must" property for any set of axioms. It precludes contradictory statements within the set of axioms itself and, consequently, contradictory theorems in the body of theorems derived from the axioms. We must remember that it is not uncommon to find contradictory theorems derived from two different sets of axioms. This is bound to happen if one of the two sets of axioms contains a statement contradictory to a statement or combination of statements in the other set of axioms. This is exactly the situation we find in theorems obtained from using, for one set of axioms, those axioms for Euclidean geometry and, for the other set of axioms, those for one of the non-Euclidean geometries. A theorem of Euclidean plane geometry says that the sum of the measures of the angles of a triangle is 180; a theorem from one of the non-Euclidean geometries says that the sum of the measures of the angles of a triangle is less than 180. Consistency, however, is a term used with reference to a particular set of axioms. Our attention is fixed on that one set, and we want to determine consistency within that set alone.

Exercise

Test the consistency of the following set of axioms:

Axiom 1. Each ogg is on an uff.
Axiom 2. No ogg is on an uff by itself.
Axiom 3. There are exactly four oggs.
Axiom 4. Not all oggs are on the same uff.
Axiom 5. There are exactly six uffs.
Axiom 6. No ogg is on more than three uffs.

Theorems Derived from a Set of Axioms

Let us now turn to the discovery and proofs of theorems derived from a set of axioms. We shall use a new set called axiom set E:

Axiom 1. Each ogg is on an uff.
Axiom 2. Not all oggs are on the same uff.
Axiom 3. No ogg is on an uff by itself.
Axiom 4. There are exactly four oggs.

If we interpret an ogg to be a dot and an uff to be an arc ($\frown$), a consistency model for axiom set E is exhibited in Figure 4-4.

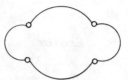

Figure 4-4

Notice that axiom 4 says that there must be exactly four oggs. This must be true in *every* consistency model for axiom set E. In the model of Figure 4-4 there are also exactly four uffs. Does this mean that there are exactly four uffs in *every* consistency model? By axiom 1 we know that there is at least one uff. Is there a theorem which says: There are exactly four uffs? We can give a consistency model which shows that the above statement does *not* follow from the axioms and hence is *not* a theorem. Such a consistency model is shown in Figure 4-5.

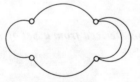

Figure 4-5

All the axioms are satisfied in Figure 4-5 and there are five uffs in this model. You can see that we could have six, seven, eight, or any greater number of uffs, just by adding more arcs as we did to Figure 4-4 to obtain Figure 4-5.

The following statement actually *is* a theorem, and we shall so label it and give the proof. You will notice that the proof follows the pattern of proofs in geometry. That is, we give a reason for each step in the proof. Each reason is based on an axiom.

Theorem. There are at least two uffs.

Proof. By axiom 4, there are exactly four oggs. Call them A, B, C, D.
By axiom 1, ogg A is on an uff.
By axiom 3, ogg A is on an uff with another ogg, say B.
Thus we have *one* uff AB.
By axiom 2, not all oggs are on the same uff, so there must be an ogg, say ogg C, not on uff AB.
By axiom 1, C is on an uff.
By axiom 3, C is on an uff with another ogg. This other ogg might be A. In this case we would have a second uff CA. This other ogg might be B, in which case we would have a second uff CB. This other uff might be neither A nor B, in which case it would have to be D. Then we would have a second uff CD. In every case, then, we have found a second uff. The theorem is thus proved. There are at least two uffs.

It is worth noticing that the proof given for this theorem used no diagrams, only words. It was not necessary to use any diagrams. In fact, many times students trying to prove theorems in geometry depend too much on diagrams. They are apt to think that certain statements are true because they look that way in the diagram. Again, the diagrams may illustrate special cases of the theorem and lead the student to assume statements that would not be true in general. One of the reasons we use oggs and uffs is to prevent any special meanings from being attached to them. The proof we used was "abstract" in the sense that we used no physical models in it.

We shall not go farther in obtaining theorems from sets A through E. They are hard to come by because we have so little to work with. We trust that enough has been done to give the flavor of theorem seeking and theorem proving.

Group Theory and the Postulational Method

Carl H. Denbow and Victor Goedicke

1. The Search for Generality

Scientists and mathematicians are forever searching for more "inclusive" analyses of the things they study. They are never so happy as when they have discovered a relationship between two kinds of phenomena formerly considered independent. An excellent example of this is to be found in Newton's discovery of the law of universal gravitation, which showed that the forces with which objects at the surface of the earth are pulled downward are exactly the same kinds of forces as those which pull the moon toward the earth and the earth toward the sun, and, in fact, that all such forces throughout the universe are given by a single law so simple that any schoolboy can learn and understand it.* One of the great recent triumphs of physics is the creation of a theory which unifies the extensive data about the behavior of electromagnetic waves and the behavior of showers of atomic particles, whereas formerly a separate theory was required for each.

In the case of the scientists, this search for more inclusive theories is easy to understand. Experience tells us that a theory which works over a wide range of

---

From pp. 117–125 in *Foundations of Mathematics* by Carl H. Denbow and Victor Goedicke. Copyright © 1959 by Carl H. Denbow and Victor Goedicke. Reprinted by permission of Harper and Row, Publishers, Inc.

* "Every object in the universe is attracted by every other body in the universe with a force which is proportional to the product of the masses and inversely proportional to the square of the distance between them."

human experience is more likely to be useful than one which works over only a narrow range. If an astronomer studying energy generation has a theory which works for all the stars, he is more likely to be on the right track than an astronomer whose theory works only for the sun.

Mathematicians also search for more inclusive theories, and for somewhat similar reasons. If we can find a single theory which unifies several branches of mathematics, we have increased the power of mathematics accordingly, because we have simplified the structure of mathematical knowledge and facilitated the task of learning and applying it. It would be misleading, however, if we asserted that this was the only motivation for the discovery of these theories. In fact, the motivation is actually esthetic as well as utilitarian. To discover a kind of hidden master pattern in the structure of mathematical knowledge is an esthetic achievement, and the study of it should give you esthetic pleasure as well as enlightenment. It is with this hope that we proceed to a discussion of one of the broad, inclusive branches of mathematics. It is called "group theory."

2. The Furniture Mover's Story

The theory of groups is rather recent in the history of mathematics. It was in fact the branch of mathematics developed by a French boy named Evaristo Galois, who died in 1832 at the age of 20. The circumstances of his death are unusual. He died in a duel which, it is believed, was rigged by persons who felt it necessary to do away with him on political grounds. His behavior indicates that he knew that he was to die. He spent the entire night before the duel writing a kind of scientific testament, feverishly committing to paper the mathematical ideas which he would have liked to develop at his leisure and present to mankind in finished form. Then he went out to face his death, and mankind was deprived of one of its truly great minds.*

Our purpose in introducing the theory of groups is to simplify several areas of mathematics by demonstrating an underlying unity which they possess. (It is interesting to note that this underlying unity goes beyond mathematics, and that other examples of group theory are to be found in art, in the theory of crystal structure, in

---

* For an interesting account of the life of Galois, see *Whom the Gods Love*, by Leopold Infeld, McGraw-Hill, 1948.

the theory of relativity, and in other unexpected areas.) But for your first introduction to group theory we have chosen an example which is very simple and direct, and nonmathematical in the ordinary sense. We hope that this illustration will show you that mathematics need not deal with numbers or quantities or geometry, and we hope further that it will show you how new kinds of mathematics can arise out of new ways of looking at familiar situations. And so we will ask you to follow us in a pathetic little story of a man whose wife was addicted to furniture moving.

Now of course most women move furniture around occasionally, in an effort to give the room "that added something." But most women, after having experimented with the same unsuccessful arrangement several times, will give it up. The wife of our hero, Mr. Adamson, was different. She could not resist the hope that if she tried a given arrangement just one more time she might be able to make it work out.

Imagine then an ordinary living room, containing among other things a sofa, a radio, and a table (Figure 4-6). At the end of each hard day's work, Mr. Adamson comes home to find that his wife has decided to try a new arrangement of these three objects. She makes her "orders of the day" explicit by writing a note, which is always

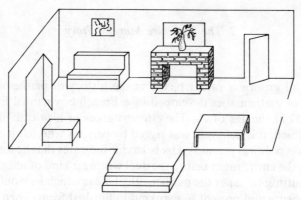

Figure 4-6. The Adamsons' Living Room

in the same form. A typical example is $\begin{bmatrix} RST \\ TRS \end{bmatrix}$. Mr. Adamson knows all too well what this means. "Put the table where the radio now is, and put the radio where the sofa now is, and put the sofa where the table now is." Mr. Adamson has been the recipient of so many of these peremptory little messages that he has named them. He calls this one G, because it was first handed to him on green paper.

Mr. Adamson early learned that he is expected to replace the objects named in the top row by those in the bottom row, and not vice versa. The top symbols represent

Group Theory and the Postulational Method

"before" and the bottom ones "after." He also noticed that there are several messages which look different but which contain the same instructions. For example, the message $\begin{bmatrix} SRT \\ RTS \end{bmatrix}$ contains exactly the same instructions as the one he originally received on green paper, so he calls this one G also. In fact, when he tries to tabulate all possible messages, he finds only six basically different ones. One of these he looks at with special longing; it is the message $\begin{bmatrix} RST \\ RST \end{bmatrix}$, which would require no work. Though he has never been granted this boon, he has included it for the sake of making his list complete, and has named it E (for easy). The message $\begin{bmatrix} RST \\ STR \end{bmatrix}$ he has named J, because he can still remember the hot June day when he would have liked to go fishing instead of moving furniture. The remaining ones he named alphabetically; he used A for $\begin{bmatrix} RST \\ RTS \end{bmatrix}$ (in which the radio is not moved), and B for $\begin{bmatrix} RST \\ TSR \end{bmatrix}$, and C for $\begin{bmatrix} RST \\ SRT \end{bmatrix}$ (Table 5.1).

Table 5.1. The Six Messages

$$E = \begin{bmatrix} RST \\ RST \end{bmatrix} \qquad A = \begin{bmatrix} RST \\ RTS \end{bmatrix}$$

$$C = \begin{bmatrix} RST \\ SRT \end{bmatrix} \qquad J = \begin{bmatrix} RST \\ STR \end{bmatrix}$$

$$G = \begin{bmatrix} RST \\ TRS \end{bmatrix} \qquad B = \begin{bmatrix} RST \\ TSR \end{bmatrix}$$

Now these names, G, J, and the rest, are only names, and you may be inclined to think they are not very important. A name is only a name; it moves no furniture. But by now you should be suspicious of this agnostic attitude. We have seen that a set of language agreements is an important part of establishing a new system of thought, and that an apt naming system frequently clarifies our thinking and leads to new insights. (If this were not so, mathematics would still be a branch of English!) So let us see what will come out of this system of names which Mr. Adamson gave to his wife's repeated orders.

As Mr. Adamson now recalls it, his interest in the implications of this little game first occurred to him when his wife's aunt became ill. One night Mr. Adamson came home and found the inevitable message (this time it was *A*) accompanied by a note which said "I am spending the night with poor Aunt Susan. Your dinner is in the refrigerator." Mr. Adamson was especially hot and tired that day, and decided to have a beer and read the paper before he moved the furniture, and, human nature being what it is, when the next evening came the furniture was still unmoved and another message (*B*) has arrived. Instead of actually moving the furniture twice, he decided to find out by means of pencil and paper what the final positions of the furniture should be after the two manipulations. He wrote an R, and S, and a T in three circles to indicate the original positions of the three pieces of furniture. Underneath he wrote the names of the furniture which would replace these after manipulation *A*, and under these a third set of letters to show the results of manipulation *B*. This "lazy man's method" of moving furniture is shown in Figure 4-7.

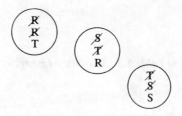

Figure 4-7

Mr. Adamson then inspected his list of the six possible manipulations to see which one would accomplish this all in one step, and discovered that *G* will do it. He abbreviated this information by writing "$A + B = G$." He proceeded happily to perform the rearrangement *G*, quite pleased with himself for having telescoped two days' messages into one. From this time onward (so depraved was his character) he shirked his work and let the messages pile up until his wife was due to return home, at which time he added the accumulated messages. He found that, no matter what messages he had received, there was always one *single* manipulation which would produce the same furniture arrangement as the entire set of manipulations. (In the language of mathematics, he had discovered that his system of manipulations had the property of closure.)

After this discovery, he sought to avoid even the work of performing the additions (many of mankind's greatest achievements, he told himself, are the result of an

Group Theory and the Postulational Method

effort to avoid work) by simply recording all his sums for future use, like an addition table. And finally (as often happens) this effort to avoid work was his undoing. His wife left him message B one Monday night, and left for Aunt Susan's; on Tuesday night she sent over message A; he recalled that he had already proved that $A + B = G$. He was just completing manipulation G on the furniture when his wife arrived and asked him, in icy tones, just what he thought he was doing. Flushed and unhappy, he worked out the separate messages on paper and found that message B followed by message A (which was the sequence his wife had ordered) gave a result different from that of message A followed by message B; in fact $A + B = G$, while $B + A = J$! In this way he discovered that addition of these messages is not commutative.

Problems

Compute the following sums.

1. $B + C$.
2. $C + B$.
3. $C + A$.
4. $A + C$.
5. $G + J$.
6. $J + G$.
7. $B + B$.

In the light of this discovery, Mr. Adamson saw the wisdom of working out a complete addition table. His results are shown in Table 4-1.

Table 4-1. Addition Table

		\multicolumn{6}{c}{Second Term}

First Term		E	A	C	J	G	B
	E	E	A	C	J	G	B
	A	A	E	J	C	B	G
	C	C	G	E	B	A	J
	J	J	B	A	G	E	C
	G	G	C	B	E	J	A
	B	B	J	G	A	C	E

By this time he began to see that he was dealing with a kind of mathematics in which the things to be added were not *numbers*, but *operations*. He also saw that his addition table could be used for any situation in which the operations were similar to his furniture moving, whether the objects rearranged were pieces of furniture, or moons of Jupiter, or carbon atoms in a molecule. In other words, he was concerned with the study of certain operations rather than the nature of the objects being operated upon. (This same accusation has sometimes been leveled at surgeons.) The addition of these operations (or elements, as he sometimes called them) has some feature in common with the addition of numbers; for example, the element E plays a role similar to that of zero in number addition. (In number addition, 2 plus zero gives 2 again, 3 plus zero equals 3, and so forth; while in Mr. Adamson's addition of elements A plus E gives A again, B plus E equals B, and so forth.) From then on he always called E "the neutral element."

Mr. Adamson's next discovery came as a result of Aunt Susan's taking a turn for the worse. (Individual welfare frequently has to be sacrificed for the advancement of science.) There came a day when Mr. Adamson found himself with an accumulation of not two, but three messages, so that he had to perform the computation $B + J + G$. He noted at once that his addition table permitted him to add only two elements at a time, and that this gave him a choice of procedure. If he added the $B + J$ first, he would obtain A, and if he then, by consulting Table 4-1, performed the addition $A + G$, he would obtain B. By using parentheses he summarized these results as follows:

$$(B + J) + G = A + G = B,$$

and since this is exactly the grouping which the problem required (after all, if he had been doing his job properly he would, at the end of the second day, have performed $B + J$, and he would then on the third day have performed G), therefore B was undoubtedly the right answer. But Mr. Adamson was now firmly held in the grip of scientific curiosity, and he could not help wondering what *would* have happened *if* he had combined the elements J and G first. Having an experimental turn of mind, he tried it. He found from Table 4-1 that $J + G = E$, so that

$$B + (J + G) = B + E = B,$$

and found that, in this case at least, the order in which he carried out the simplification makes no difference. In other words, $(B + J) + G = B + (J + G)$. Further experimentation showed him that this is always true. This result can be formulated as follows:

$$(x + y) + z = x + (y + z),$$

which is the same as the associative law of addition of numbers.

When Mr. Adamson had assembled his addition table, he noted with longing that there were several pairs which gave E as a sum, so that if his luck were good his wife might give him two successive messages which totaled E, and thus save him any work whatever. For example, $G + J = E$; or in other words, J is the element which undoes element G, and leaves the original furniture arrangement as if no manipulation had been performed at all. For this reason we say that J is the "inverse" of G. Similarly, we note that G is the inverse of J, since $J + G$ also equals E. We note from Table 4-1 that A is its own inverse since $A + A = E$.

Problem

8. Find the inverse of each of the six elements.

The more Mr. Adamson studied the table, the more he was impressed by the differences between addition of these elements and addition of numbers. He saw that writing $G + J = E$ is something like writing $7 + (-7) = 0$; in words, J can be regarded as something like the negative of G. But then what of the equation $A + A = E$? Here is an element which is its own "negative"; if you add it to itself it gives zero! And what about the equation $G + G = J$ and $J + J = G$? Here are two elements such that adding the first to itself gives the second, while adding the second to itself gives the first! He saw that he would have to be very careful to avoid carrying preconceptions from one system of mathematics to another.

Mr. Adamson often speculated about the further possibilities of this new system of mathematics. He knew, for example, that in the mathematics of numbers you can always solve an equation like $x + 3 = 7$, and that the solution is unique. That is, if $x + 3 = 7$, then there is one and only one number which x can be, namely 4. Now suppose that you wish to find an element to which you could add A to obtain J; in other words, suppose you wish to solve the equation $X + A = J$. Would such an equation always have a solution, and would the solution always be unique? These questions remained unsolved for many weeks, until his attention was forcibly directed to them in the following way. He had added the Monday and Tuesday messages and found that their sum was J; then he received a note on Wednesday which read, "I have changed my mind; don't carry out the Tuesday instructions but only the Monday instructions." But by this time he had discarded the original messages, and he remembered only that the Tuesday message was A and the total of the two was J! Necessity is the mother of invention, so he set up the equation $X + A = J$, where X was the forgotten Monday message. Then he solved the equation by the simple expedient of locating the *column* headed A, and running his finger down it until he came to J, then finding the number at the left end of this row, which turned out to be B. In other words, he found by a systematic search of Table 5.2 that $B + A = J$; and B was therefore the answer to his problem. He noticed that the answer was unique (since J occurred only once in the column) and that in fact every such equation had one and only one solution, since every element occurs once and only once in each column and in each row.

But Mr. Adamson's curiosity had been aroused by the effort to solve equations in this system. He wondered if the equation could not be solved by some method neater than systematic search, just as equations in arithmetic or algebra can be solved. To solve the equation

$$x + 3 = 7,$$

he recalled, it is necessary to subtract 3 (or to add "minus" 3) on each side of the equation, giving

$$x + 3 + (-3) = 7 + (-3),$$

which, when simplified, yields

$$x = 4.$$

By analogy, we should expect to add the inverse of A to each side of the equation

$$X + A = J,$$

since the inverse is in a sense the negative of an element. Here we must remember that in our system pre-adding is different from post-adding; if for instance we pre-add A to B we get $A + B$, which is equal to G, but if we post-add A to B we get $B + A$, which is equal to J. Let us post-add the inverse of A (which happens to be A itself) to both sides of the equation. (Obviously if we post-add on one side of an equation we must post-add on the other side.) This gives us

$$X + A + A = J + A,$$

and since A is its own inverse, we can replace $A + A$ by E:

$$X + E = J + A.$$

From the tables, we see that $X + E = X$, for every X, and $J + A = B$. Making these substitutions, we obtain finally

$$X = B.$$

Thus Mr. Adamson found that equations can be solved by manipulation, just as in the algebra of numbers.

Editorial note: I have omitted from the above selection something rather important: the definition of a group. However, the article contains all the information you need to create the definition yourself. Try to construct a definition of "group" that uses four characterizing axioms. It may be helpful to proceed as follows: A group G is a mathematical system consisting of a set, which we will also call G, and an operation defined for pairs of things in G. The operation denoted in the article by "+" satisfies the following axioms.

1. _____
2. _____
3. _____
4. _____

Part Five

Logic and Proof

Introduction

The next two articles deals with what can happen when big mathematical guns are used on certain types of questions. We have seen that man from humble beginnings devised the axiomatic approach to problems. He felt that now he would have a way to obtain the truth. He discovered in his axiomatics that insisting on "real world" starting points and interpretations got him into considerable trouble. He dropped content. This did not solve all his problems, as these articles will show. Even man's sophistication in logic probably cannot lead him to truth.

These articles do not merely raise questions; they also suggest possible helps. I did not say solutions, for in certain respects no solutions exist. At best, we can hope for remedies or new frames of reference.

Paradox

W. V. Quine

Frederic, the young protagonist of *The Pirates of Penzance*, has reached the age of 21 after passing only five birthdays. Several circumstances conspire to make this possible. Age is reckoned in elapsed time, whereas a birthday has to match the date of birth; and February 29 comes less frequently than once a year.

Granted that Frederic's situation is possible, wherein is it paradoxical? Merely in its initial air of absurdity. The likelihood that a man will be more than n years old

From *Scientific American*, April 1962. Reprinted with permission. Copyright © 1962 by Scientific American, Inc. All rights reserved.

on his nth birthday is as little as one to 1,460 or slightly better if we allow for seasonal trends; and this likelihood is so slight that we easily forget its existence.

May we say in general, then, that a paradox is just any conclusion that at first sounds absurd but that has an argument to sustain it? In the end I think this account stands up pretty well. But it leaves much unsaid. The argument that sustains a paradox may expose the absurdity of a buried premise or of some preconception previously reckoned as central to physical theory, to mathematics or to the thinking process. Catastrophe may lurk, therefore, in the most innocent-seeming paradox. More than once in history the discovery of paradox has been the occasion for major reconstruction at the foundations of thought. For some decades, indeed, studies of the foundation of mathematics have been confounded and greatly stimulated by confrontation with two paradoxes, one propounded by Bertrand Russell in 1901 and the other by Kurt Gödel in 1931.

As a first step onto this dangerous ground, let us consider another paradox: that of the village barber. This is not Russell's great paradox of 1901, to which we shall come, but a lesser one that Russell attributed to an unnamed source in 1918. In a certain village there is a man, so the paradox runs, who is a barber; this barber shaves all and only those men in the village who do not shave themselves. Query: Does the barber shave himself?

Any man in this village is shaved by the barber if and only if he is not shaved by himself. Therefore in particular the barber shaves himself if and only if he does not. We are in trouble if we say the barber shaves himself and we are in trouble if we say he does not.

Now compare the two paradoxes. Frederic's situation seemed absurd at first, but a simple argument sufficed to make us acquiesce in it for good. In the case of the barber, on the other hand, the conclusion is too absurd to acquiesce in at any time.

What we are to say to the argument that goes to prove this unacceptable conclusion? Happily it rests on assumptions. We are asked to swallow a story about a village and a man in it who shaves all and only those men in the village who do not shave themselves. This is the source of our trouble; grant this and we end up saying, absurdly, that the barber shaves himself if and only if he does not. The proper conclusion to draw is just that there is no such barber. We are confronted with nothing more mysterious than what logicians have been referring to for a couple of thousand years as a *reductio ad absurdum*. We disprove the barber by assuming him and deducing the absurdity that he shaves himself if and only if he does not. The paradox is simply a proof that no village can contain a man who shaves all and only those men in it who do not shave themselves. This sweeping denial at first sounds absurd; why should there not be such a man in a village? But the argument shows why not, and so we acquiesce in the sweeping denial just as we acquiesced in the possibility, absurd on first exposure, of Frederic's being so much more than five years old on his fifth birthday.

Both paradoxes are alike, after all, in sustaining prima facie absurdities by conclusive argument. What is strange but true in the one paradox is that one can be $4n$ years old on one's nth birthday; what is strange but true in the other paradox is that no village can contain a man who shaves all and only those men in the village who do not shave themselves.

Still, I would not limit the word "paradox" to cases where what is purportedly established is true. I shall call these, more particularly, veridical, or truth-telling, paradoxes. For the name of paradox is suited equally to falsidical ones. (This word is not so barbarous as it sounds; *falsidicus* occurs twice in Plautus and twice in earlier writers.)

The Frederic paradox is a veridical one if we take its proposition not as something about Frederic but as the abstract truth that a man can be $4n$ years old on his nth birthday. Similarly, the barber paradox is a veridical one if we take its proposition as being that no village contains such a barber. A falsidical paradox, on the other hand, is one whose proposition not only seems at first absurd but also is false, there being a fallacy in the purported proof. Typical falsidical paradoxes are the comic misproofs that $2 = 1$. Most of us have heard one or another such. Here is the version offered by the 19th-century English mathematician Augustus De Morgan: Let $x = 1$. Then $x^2 = x$. So $x^2 - 1 = x - 1$. Dividing both sides by $x - 1$, we conclude that $x + 1 = 1$; that is, since $x = 1$, $2 = 1$. The fallacy comes in the division by $x - 1$, which is 0.

Instead of "falsidical paradox" could I say simply "fallacy"? Not quite. Fallacies can lead to true conclusions as well as false ones, and to unsurprising conclusions as well as surprising ones. In a falsidical paradox there is always a fallacy in the argument, but the proposition purportedly established has furthermore to seem absurd and to be indeed false.

Some of the ancient paradoxes of Zeno belong under the head of falsidical paradoxes. Take the one about Achilles and the tortoise. Generalized beyond these two fictitious characters, what the paradox purports to establish is the absurd proposition that so long as a runner keeps running, however slowly, another runner can never overtake him. The argument is that each time the pursuer reaches a spot where the pursued has been, the pursued has moved a bit beyond. When we try to make this argument more explicit, the fallacy that emerges is the mistaken notion that any infinite succession of intervals of time has to add up to all eternity. Actually when an infinite succession of intervals of time is so chosen that the succeeding intervals become shorter and shorter, the whole succession may take either a finite or an infinite time. It is a question of a convergent series.

Grelling's Paradox

The realm of paradox is not clearly exhausted even by the veridical and falsidical paradoxes together. The most startling of all paradoxes are not clearly assignable to either of these domains. Consider the paradox, devised by the German mathematician Kurt Grelling in 1908, concerning the heterological, or nonself-descriptive, adjectives.

To explain this paradox requires first a definition of the autological, or self-descriptive, adjective. The adjective "short" is short; the adjective "English" is English; the adjective "adjectival" is adjectival; the adjective "polysyllabic" is polysyllabic. Each of these adjectives is, in Grelling's terminology, autological: each is true of itself. Other adjectives are heterological; thus "long," which is not a long adjective; "German," which is not a German adjective; "monosyllabic," which is not a monosyllabic one.

Grelling's paradox arises from the query: Is the adjective "heterological" an autological or a heterological one? We are as badly off here as we were with the barber. If we decide that "heterological" is autological, then the adjective is true of itself. But that makes it heterological rather than autological, since whatever the adjective "heterological" is true of is heterological. If we therefore decide that the adjective "heterological" is heterological, then it is true of itself, and that makes it autological.

Our recourse in a comparable quandary over the village barber was to declare a *reductio ad absurdum* and conclude that there was no such barber. Here, however, there is no interim premise to disavow. We merely define the adjective "heterological" and asked if it was heterological. In fact, we can get the paradox just as well without the adjective and its definition. "Heterological" was defined as meaning "not true of self"; we can therefore ask if the adjectival phrase "not true of self" is true of itself. We find that it is if and only if it is not, hence that it is and it is not; and so we have our paradox.

Thus viewed, Grelling's paradox seems unequivocally falsidical. Its proposition is a self-contradictory compound proposition to the effect that our adjective is and is not true of itself. But this paradox contrasts strangely with the falsidical paradoxes of Zeno, or of $2 = 1$, in that we are at a loss to spot the fallacy in the argument. It may for this reason be best seen as representing a third class of paradoxes, separate from the veridical and falsidical ones.

Antinomies

The paradoxes of this class are called antinomies, and it is they that bring on the crises in thought. An antinomy produces a self-contradiction by accepted ways of reasoning. It establishes that some tacit and tructed pattern of reasoning must be made explicit and henceforward be avoided or revised.

Take Grelling's paradox, in the form in which it shows the adjective phrase "not true of self" to be both true and false of itself. What tacit principles of reasoning does the argument depend on? Notably this one, the adjective "red" is true of a thing if and only if the thing is red: the adjective "big" is true of a thing if and only if the thing is big; the adjective "not true of self" is true of a thing if and only if the thing is not true of itself; and so on. This last case of the principle is the case that issues directly in the paradox.

There is no denying that this principle is constantly used, tacitly, when we speak of adjectives as true of things: the adjective "red" is true of a thing if and only if the thing is red, and correspondingly for all adjectives. This principle simply reflects what we mean in saying that adjectives are true of things. It is a hard principle to distrust, and yet it is obviously the principle that is to blame for our antinomy. The antinomy is directly a case of this principle. Take the adjective in the principle as the adjectival phrase "not true of self" instead of the adjective "red," and take the "thing" in the principle, of which the adjective is to be true, as that adjective over again; thereupon the principle says outright that "not true of self" is true of itself if and only if it is not true of itself. So the principle must be abandoned or at least somehow restricted.

Yet so faithfully does the principle reflect what we mean in calling adjectives true of things that we cannot abandon it without abjuring the very expression "true of" as pernicious nonsense. We could still go on using the adjectives themselves that had been said to be true of things; we could go on attributing them to things as usual; what we would be cutting out in "true of" is merely a special locution for talking about the attribution of the adjectives to the things.

This special locution, however, has its conveniences, and it would be missed. In fact, we do not have to do without it altogether. After all, to speak of adjectives as true or not true of things makes trouble only in a special case, involving one special adjective, namely the phrase "not true of self," in attribution to one special thing, namely that some phrase over again. If we forswear the use of the locution "true of" in connection with this particular phrase in relation to itself as object, we thereby silence our antinomy and may go on blithely using the locution "true of" in other cases as always, pending the discovery of further antinomies.

Actually related antinomies are still forthcoming. To inactivate the lot we have to cut a little deeper than our one case; we have to forswear the use of "true of" not only in connection with "not true of self" but also in connection with various other phrases relating to truth and in such connections we have to forswear the use of various other truth locutions. First let us look at some of the antinomies that would otherwise threaten.

The Paradox of Epimenides

There is the ancient paradox of Epimenides the Cretan, who said that all Cretans were liars. If he spoke the truth, he was a liar. It seems that the paradox may have reached the ears of St. Paul and that he missed the point of it. He warned, in his epistle to Titus: "One of themselves, even a prophet of their own, said, The Cretans are always liars."

Actually the paradox of Epimenides is untidy; there are loopholes. Perhaps some Cretans were liars, notably Epimenides, and others were not; perhaps Epimenides was a liar who occasionally told the truth; either way it turns out that the contradiction vanishes. Something of paradox can be salvaged with a little tinkering; but we do better to switch to a different and simpler rendering, also ancient, of the same idea. This is the *pseudomenon*, which runs simply: "I am lying." We can even drop the indirectness of a personal reference and speak directly of the sentence: "This sentence is false." Here we seem to have the irreducible essence of antinomy: a sentence that is true if and only if it is false.

In an effort to clear up this antinomy it has been protested that the phrase "This sentence," so used, refers to nothing. This is claimed on the ground that you cannot get rid of the phrase by supplying a sentence that is referred to. For what sentence does the phrase refer to? The sentence "This sentence is false." If, accordingly, we supplant the phrase. "This sentence" by a quotation of the sentence referred to, we get: " 'This sentence is false' is false." But the whole outside sentence here attributes falsity no longer to itself but merely to something other than itself, thereby engendering no paradox.

If, however, in our perversity we are still bent on constructing a sentence that does attribute falsity unequivocally to itself, we can do so thus: "'Yields a falsehood when appended to its own quotation' yields a falsehood when appended to its own quotation." This sentence specifies a string of nine words and says of this string that if you put down twice, with quotation marks around the first of the two occurrences, the result is false. But that result is the very sentence that is doing the telling. The sentence is true if and only if it is false, and we have our antinomy.

This is a genuine antinomy, on a par with the one about "heterological," or "false of self," or "not true of self," being true of itself. But whereas that earlier one turned on "true of," through the construct "not true of self," this new one turns merely on "true," through the construct "falsehood," or "statement not true." We can avoid both antinomies, and others related to them, by ceasing to use "true of" and "true" and their equivalents and derivatives, or at any rate ceasing to apply such truth locutions to adjectives or sentences that themselves contain such truth locutions.

This restriction can be relaxed somewhat by admitting a hierarchy of truth locutions, as suggested by the work of Bertrand Russell and the Polish mathematician Alfred Tarski, who is now at the University of California. The expressions "true," "true of," "false" and related ones can be used with numerical subscripts "0," "1," "2," and so on always attached or imagined: thus "$true_0$," "$true_1$," "$true_2$," "$false_0$" and so on. Then we can avoid the antinomies by taking care, when a truth locution (T) is applied to a sentence or other expression (S), that the subscript on T is higher than any subscript inside S. Violations of this restriction would be treated as meaningless, or ungrammatical, rather than as true or false sentences. For instance, we could meaningfully ask whether the adjectives "long" and "short" are $true_0$ of themselves, the answers are respectively no and yes. But we could not meaningfully speak of the phrase "not $true_0$ of self" as $true_0$ or $false_0$ of itself, we would have to ask whether it is $true_1$ or $false_1$ of itself, and this is a question that leads to no antinomy. Either way the question can be answered with a simple and unpenalized negative.

This point deserves to be restated: Whereas "long" and "short" are adjectives that can meaningfully be applied to themselves, falsely in the one case and truly in the other, on the other hand "$true_0$ of self" and "not $true_0$ of self" are adjectival phrases that cannot be applied to themselves meaningfully at all, truly or falsely. Therefore to the question "Is '$true_0$ of self' $true_1$ of itself?" the answer is no, the adjectival phrase "$true_0$ of itself" is meaningless of itself rather than $true_1$ of itself.

Next let us consider, in terms of subscripts, the most perverse version of the *pseudomenon*. We have now, for meaningfulness, to insert subscripts on the two occurrences of the word "falsehood," and in ascending order, thus: " 'Yields a $falsehood_0$ when appended to its own quotation yields a $falsehood_1$ when appended to its own quotation." Thereupon paradox vanishes. This sentence is unequivocally false. What it tells us is that a certain described form of words is $false_1$, namely the form of words: " 'Yields a $falsehood_0$ when appended to its own quotation' yields a $falsehood_0$ when appended to its own quotation." But in fact this form of words is not false$_1$; it is meaningless. So the preceding sentence, which said that this form of words was false$_1$, is false. It is false$_2$.

This may seem an extravagant way of eliminating antinomies. But it would be much more costly to drop the word "true," and related locutions, once and for all. At an intermediate cost one could merely leave off applying such locutions to expressions containing such locutions. Either method is less economical than this method of subscripts. The subscripts do enable us to apply truth locutions to expressions containing such locutions, although in a manner disconcertingly at variance with custom. Each resort is desperate; each is an artificial departure from natural and established usage. Such is the way of antinomies.

A veridical paradox packs a surprise, but the surprise quickly dissipates itself as we ponder the proof. A falsidical paradox packs a surprise, but it is seen as a false alarm when we solve the underlying fallacy. An antinomy, however, packs a surprise that can be accommodated by nothing less than a repudiation of part of our conceptual heritage.

Revision of a conceptual scheme is not unprecedented. It happens in a small way with each advance in science, and it happens in a big way with the big advances, such as the Copernican revolution and the shift from Newtonian mechanics to Einstein's theory of relativity. We can hope in time even to get used to the biggest such changes and to find the new schemes natural. There was a time when the doctrine that the earth revolves around the sun was called the Copernican paradox, even by the men who accepted it. And perhaps a time will come when truth locutions without implicit subscripts, or like safeguards, will really sound as nonsensical as the antinomies show them to be.

Conversely, the falsidical paradoxes of Zeno must have been, in his day, genuine antinomies. We in our latter-day smugness point to a fallacy: the notion that an infinite succession of intervals must add up to an infinite interval. But surely this was part and parcel of the conceptual scheme of Zeno's day. Our recognition of convergent series, in which an infinite number of segments add up to a finite segment, is from Zeno's vantage point an artificiality comparable to our new subscripts on truth locutions. Perhaps these subscripts will seem as natural to our descendants of A.D. 4000, granted the tenuous hypothesis of there being any, as the convergent series does to us. One man's antinomy is another man's falsidical paradox, give or take a couple of thousand years.

I have not, by the way, exhausted the store of latter-day antinomies. Another good one is attributed by Russell to a librarian named Berry. Here the theme is numbers and syllables. Ten has a one-syllable name. Seventy-seven has a five-syllable name. The seventh power of seven hundred seventy-seven has a name that, if we were to work it out, might run to 100 syllables or so; but this number can also be specified more briefly in other terms. I have just specified it in 15 syllables. We can be sure, however, that there are no end of numbers that resist all specification, by name or description, under 19 syllables. There is only a finite stock of syllables altogether, and hence only a finite number of names or phrases of less than 19 syllables, whereas there are an infinite number of positive integers. Very well, then; of those numbers not specifiable in less than 19 syllables, there must be a least. And here is our antinomy: the least number not specifiable in less than nineteen syllables is specifiable in 18 syllables. I have just so specified it.

This antinomy belongs to the same family as the antinomies that have gone before. For the key word of this antinomy, "specifiable," is interdefinable with "true of." It is one more of the truth locutions that would take on subscripts under the Russell–Tarski plan. The least number not specifiable$_0$ in less than nineteen syllables is indeed specifiable$_1$ in 18 syllables, but it is not specifiable$_0$ in less than 19 syllables; for all I know it is not specifible$_0$ in less than 23. This resolution of Berry's antinomy is the one that would come through automatically if we paraphrase "specifiable" in terms of "true of" and then subject "true of" to the subscript treatment.

Russell's Antinomy

Not all antinomies belong to this family. The most celebrated of all antinomies, discovered by Russell in 1901, belongs outside this family. It has to do with self-membership of classes. Some classes are members of themselves; some are not. For example, the class of all classes that have more than five members clearly has more than five classes as members; therefore the class is a member of itself. On the other hand, the class of all men is not a member of itself, not being a man. What of the class of all classes that are not members of themselves? Since its members are the nonself-members, it qualifies as a member of itself if and only if it is not. It is and it is not: antinomy's by now familiar face.

Russell's antinomy bears a conspicuous analogy to Grelling's antinomy of "not true of self," which it long antedates. But Russell's antinomy does not belong to the same family as the Epimenides antinomy and those of Berry and Grelling. By this I mean that Russell's antinomy cannot be blamed on any of the truth locutions, nor is it resolved by subjecting those locutions to subscripts. The crucial words in Russell's antinomy are "class" and "member," and neither of these is definable in terms of "true," "true of" or the like.

I said earlier that an antinomy establishes that some tacit and trusted pattern of reasoning must be made explicit and be henceforward avoided or revised. In the case of Russell's antinomy, the tacit and trusted pattern of reasoning that is found wanting is this: for any condition you can formulate, there is a class whose members are the things meeting the condition.

This principle is not easily given up. The almost invariable way of specifying a class is by stating a necessary and sufficient condition for belonging to it. When we have stated such a condition we feel that we have "given" the class and can scarcely make sense of there not being such a class. The class may be empty, yes; but how could there not be such a class at all? What substance can be asked for it that the membership condition does not provide? Yet such exhortations avail us nothing in the face of the antinomy, which simply proves the principle untenable. It is a simple point of logic, once we look at it, that there is no class, empty or otherwise, that has as members precisely the classes that are not members of themselves. It would have to have itself as member if and only if it did not.

Russell's antinomy came as a shock to Gottlob Frege, the German mathematician who founded mathematical logic. In his *Grundgesetze der Arithmetik* Frege thought that he had secured the foundations of mathematics in the self-consistent laws of logic. He received a letter from Russell as the second volume of this work was on its way to press. "Arithmetic totters," Frege is said to have written in answer. An appendix that he added to the volume opens with the words: "A scientist can hardly encounter anything more undesirable than to have the foundation collapse just as the work is finished. I was put in this position by a letter from Bertrand Russell . . .".

In Russell's antinomy there is more than a hint of the paradox of the barber. The parallel is, in truth, exact. It was a simple point of logic that there was in no village a man who shaved all and only those men in the village who did not shave themselves; he would shave himself if and only if he did not. The barber paradox was a veridical paradox showing that there is no such barber. Why is Russell's antinomy then not a veridical paradox showing that there is no class whose members are all and only the nonself-members? Why does it count as an antinomy and the barber paradox not? The reason is that there has been in our habits of thought an overwhelming presumption of there being such a class but no presumption of there being such a barber. The barber paradox barely qualifies as paradox in that we are mildly surprised at being able to exclude the barber on purely logical grounds by reducing him to absurdity. Even this surprise ebbs as we review the argument; and anyway we had never positively believed in such a barber. Russell's paradox is a genuine antinomy because of the fundamental nature of the principle of class existence that it compels us to give up. When in a future century the absurdity of that principle has become a commonplace, and some substitute principle has enjoyed long enough tenure to take on somewhat the air of common sense, perhaps we can begin to see Russell's paradox as no more than a veridical paradox, showing that there is no such class as that of the nonself-members. One man's antinomy can be another man's veridical paradox, and one man's veridical paradox can be another man's platitude.

Russell's antinomy made for a more serious crisis still than did Grelling's and Berry's and the one about Epimenides. For these strike at the semantics of truth and denotation, but Russell's strikes at the mathematics of classes. Classes are appealed to in an auxiliary way in most branches of mathematics, and increasingly so as passages of mathematical reasoning are made more explicit. The basic principle of classes that is tacitly used, at virtually every turn where classes are involved at all, is precisely the class-existence principle that is described by Russell's antinomy.

I spoke of Grelling's antinomy and Berry's and the Epimenides as all in a family, to which Russell's antinomy does not belong. For its part, Russell's antinomy has family connections of its own. In fact, it is the first of an infinite series of antinomies, as follows. Russell's antinomy shows that there is no class whose members are precisely the classes that are not members of themselves. Now there is a parallel antinomy that shows there is no class whose members are precisely the classes that are not members of members of themselves. Further, there is an antinomy that shows there is no class whose members are precisely the classes that are not members of members of members of themselves. And so on ad infinitum.

All these antinomies, and other related ones, can be inactivated by limiting the guilty principle of class existence in a very simple way. The principle is that for any membership condition you can formulate there is a class whose members are solely the things meeting the condition. We get Russell's antinomy and all the others of its series by taking the condition as nonmembership in self, or nonmembership in members of self, or the like. Each time the trouble comes of taking a membership condition that itself talks in turn of membership and nonmembership. If we withhold our principle of class existence from cases where the membership condition mentions membership, Russell's antinomy and related ones are no longer forthcoming. This restriction on class existence is parallel to a restriction on the truth locutions that we contemplated for a while, before bringing in the subscripts; namely, not to apply the truth locutions to expressions containing any of the truth locutions.

Happily we can indeed withhold the principle of class existence from cases where the membership condition mentions membership, without unsettling those branches of mathematics that make only incidental use of classes. This is why it has been possible for most branches of mathematics to go on blithely using classes as auxiliary apparatus in spite of Russell's and related antinomies.

The Mathematics of Classes

There is a particular branch of mathematics in which the central concern is with classes: general set theory. In this domain one deals expressly with classes of classes, classes of classes of classes, and so on, in ways that would be paralyzed by the restriction just now contemplated: withholding the principle of class existence from cases where the membership condition mentions membership. So one tries in general set theory to manage with milder restrictions.

General set theory is rich in paradox. Even the endless series of antinomies that I mentioned above, of which Russell's was the first, by no means exhausts this vein of paradox. General set theory is primarily occupied with infinity—infinite classes, infinite numbers—and so is involved in paradoxes of the infinite. A rather tame old paradox under this head is that you can exhaust the members of a whole class by correlating them with the members of a mere part of the class. For instance, you can correlate all the positive integers with the multiples of 10, thus: 1 with 10, 2 with 20, 3 with 30 and so on. Every positive integer gets disposed of; there are as many multiples of 10 as integers altogether. This is no antinomy but a veridical paradox. Among adepts in the field it even loses the air of paradox altogether, as is indeed the way of veridical paradox.

Georg Cantor, the 19th-century pioneer in general set theory and infinite arithmetic, proved that there are always more classes of things of a given kind than there are things of that kind; more classes of cows than cows. A distinct air of paradox suffuses his proof of this.

First note the definition of "more." What it means when one says there are more things of one kind than another is that every correlation of things of the one kind to things of the other fails to exhaust the things of the one kind. So what Cantor is proving is that no correlation of cow classes to cows accommodates all the cow classes. The proof is as follows. Suppose a correlation of cow classes to cows. It can be any arbitrary correlation; a cow may or may not belong to the class correlated with it. Now consider the cows, if any, that do not belong to the classes correlated with them. These cows themselves form a cow class, empty or not. And it is a cow class that is not correlated with any cow. If the class were so correlated, that cow would have to belong to the class if and only if it did not.

This argument is typical of the arguments in general set theory that would be sacrificed if we were to withhold the principle of class existence from cases where the membership condition mentions membership. The recalcitrant cow class that clinched the proof was specified by a membership condition that mentioned membership. The condition was nonmembership in the correlated cow class.

But what I am more concerned to bring out, regarding the cow-class argument, is its air of paradox. The argument makes its negative point in much the same way that the veridical barber paradox showed there to be no such barber, and in much the same way that Russell's antinomy showed there to be no class of nonself-members. So in Cantor's theorem—a theorem not only about cows and their classes but also about things of any sort and their classes—we see paradox, or something like it, seriously at work in the advancement of theory. His theorem establishes that for every class, even every infinite class, there is a larger class: the class of its subclasses.

So far, no antinomy. But now it is a short step to one. If for every class there is a larger class, what of the class of everything? Such is Cantor's antinomy. If you review the proof of Cantor's theorem in application directly to this disastrous example—speaking therefore not of cows but of everything—you will quickly see that Cantor's antinomy boils down, after all, to Russell's.

So the central problem in laying the foundations of general set theory is to inactivate Russell's antinomy and its suite. If such theorems as Cantor's are to be kept, the antinomies must be inactivated by milder restrictions than the total withholding of the principle of class existence from cases where the membership condition mentions membership. One tempting line is a scheme of subscripts analogous to the scheme used in avoiding the antinomies of truth and denotation. Something like this line was taken by Russell himself in 1908, under the name of the theory of logical types. A very different line was proposed in the same year by the German mathematician Ernst Zermelo, and further variations have been advanced in subsequent years.

All such foundations for general set theory have as their point of departure the counsel of the antinomies; namely, that a given condition, advanced as a necessary and sufficient condition of membership in some class, may or may not really have a class corresponding to it. So the various alternative foundations for general set theory differ from one another with respect to the membership conditions to which they do and do not guarantee corresponding classes. Nonself-membership is of course a condition to which none of the theories accord corresponding classes. The same holds true for the condition of not being a member of any own member; and for the conditions that give all the further antinomies of the series that began with Russell's; and for any membership condition that would give rise to any other antinomy, if we can spot it.

But we cannot simply withhold each antinomy-producing membership condition and assume classes corresponding to the rest. The trouble is that there are membership conditions corresponding to each of which, by itself, we can innocuously assume a class, and yet these classes together can yield a contradiction. We are driven to seeking optimum consistent combinations of existence assumptions, and consequently there is a great variety of proposals for the foundations of general set theory. Each proposed scheme is unnatural, because the natural scheme is the unrestricted one that the antinomies discredit; and each has advantages, in power or simplicity or in attractive consequences in special directions, that each of its rivals lacks.

I remarked earlier that the discovery of antinomy is a crisis in the evolution of thought. In general set theory the crisis began 60 years ago and is not yet over.

Gödel's Proof

Up to now the heroes or villains of this piece have been the antinomies. Other paradoxes have paled in comparison. Other paradoxes have been less startling to us, anyway, and more readily adjusted to. Other paradoxes have not precipitated 60-year crises, at least not in our time. When any of them did in the past precipitate crises that durable (and surely the falsidical paradoxes of Zeno did), they themselves qualified as antinomies.

Let me, in closing, touch on a latter-day paradox that is by no means an antinomy but is strictly a veridical paradox, and yet is comparable to the antinomies in the pattern of its proof, in the surprisingness of the result and even in its capacity to precipitate a crisis. This is Gödel's proof of the incompletability of number theory.

What Kurt Gödel proved, in that great paper of 1931, was that no deductive system, with axioms however arbitrary, is capable of embracing among its theorems all the truths of the elementary arithmetic of positive integers unless it discredits itself by letting slip some of the falsehoods too. Gödel showed how, for any given deductive system, he could construct a sentence of elementary number theory that would be true if and only if not provable in that system. Every such system is therefore either incomplete, in that it misses a relevant truth, or else bankrupt, in that it proves a falsehood.

Gödel's proof may conveniently be related to the Epimenides paradox or the *pseudomenon* in the "yields a falsehood" version. For "falsehood" read "nontheorem," thus: " 'Yields a nontheorem when appended to its own quotation' yields a nontheorem when appended to its own quotation."

This statement no longer presents an antinomy, because it no longer says of itself that it is false. What it does say of itself is that it is not a theorem (of some deductive theory that I have not yet specified). If it is true, here is one truth that that deductive theory, whatever it is, fails to include as a theorem. If the statement is false, it is a theorem, in which event that deductive theory has a false theorem and so is discredited.

What Gödel proceeds to do, in getting his proof of the incompletability of number theory, is the following. He shows how the sort of talk that occurs in the above statement—talk of nontheoremhood and of appending things to quotations—can be mirrored systematically in arithmetical talk of integers. In this way, with much ingenuity, he gets a sentence purely in the arithmetical vocabulary of number theory that inherits that crucial property of being true if and only if not a theorem of number theory. And Gödel's trick works for any deductive system we may choose as defining "theorem of number theory."

Gödel's discovery is not an antinomy but a veridical paradox. That there can be no sound and complete deductive systematization of elementary number theory, much less of pure mathematics generally, is true. It is decidedly paradoxical, in the sense that it upsets crucial preconceptions. We used to think that mathematical truth consisted in provability.

Like any veridical paradox, this is one we can get used to, thereby gradually sapping its quality of paradox. But this one takes some sapping. And mathematical logicians are at it, most assiduously. Gödel's result started a trend of research that has grown in 30 years to the proportions of a big and busy branch of mathematics sometimes called proof theory, having to do with recursive functions and related matters, and embracing indeed a general abstract theory of machine computation. Of all the ways of paradoxes, perhaps the quaintest is their capacity on occasion to turn out to be so very much less frivolous than they look.

Questions on Paradox

1. What is a paradox?
2. Describe in your own words the Russell and Gödel paradoxes.
3. Distinguish veridical paradoxes, falsidical paradoxes, and antinomies.
4. What are truth locutions?
5. How can each of the three paradoxes be removed?

6. Explain Zeno's paradox.
7. As a special project, look up Frege and his work, and report to your class.
8. How does the Russell paradox differ from the barber paradox?

Symbolic Logic

John E. Pfeiffer

What number added to one fifth of itself equals 21? This problem was too difficult for most of the scholars of ancient Egypt. According to papyrus records, many arithmeticians struggled with it in vain before a patient Egyptian finally arrived at the correct answer about 1600 B.C. Today a ninth-grade algebra student can find the answer in a moment: $x + x/5 = 21$; therefore $x = 17\frac{1}{2}$. What made the problem hard for the Egyptians was that they lacked our handy symbols, i.e., digits for numbers and x for the unknown. Since they had to use words to represent numbers, their operations in arithmetic and algebra were cumbersome and slow.

The substitution of symbols for words is one of the things that has been largely responsible for man's progress in science. Yet in the process of logic—the basic tool with which we must test all ideas and also solve most of our everyday problems—we are still laboring under the Egyptians' handicap. We are at the mercy of the inadequacies and clumsiness of words.

From *Scientific American*, December 1950. Reprinted with permission. Copyright © 1950 by Scientific American, Inc. All rights reserved.

Consider this simple exercise in logic, taken from a textbook on the subject by Lewis Carroll, mathematician and author of *Alice's Adventures in Wonderland*:

> No kitten that loves fish is unteachable.
> No kitten without a tail will play with a gorilla.
> Kittens with whiskers always love fish.
> No teachable kitten has green eyes.
> No kittens have tails unless they have whiskers.

One, and only one, deduction can be drawn from this set of statements. After considerable trial and error you may find the answer by rewording and rearranging the statements:

> Green-eyed kittens cannot be taught.
> Kittens that cannot be taught do not love fish.
> Kittens that do not love fish have no whiskers.
> Kittens that have no whiskers have no tails.
> Kittens that have no tails will not play with a gorilla.

The one valid deduction, then, is that green-eyed kittens will not play with a gorilla.

But now take a problem that is somewhat more complicated. The following is adapted from an examination in logic prepared recently by the mathematician Walter Pitts of the Massachusetts Institute of Technology:

If a mathematician does not have to wait 20 minutes for a bus, then he either likes Mozart in the morning or whisky at night, but not both.

If a man likes whisky at night, then he either likes Mozart in the morning and does not have to wait 20 minutes for a bus or he does not like Mozart in the morning and has to wait 20 minutes for a bus or else he is no mathematician.

If a man likes Mozart in the morning and does not have to wait 20 minutes for a bus, then he likes whisky at night.

If a mathematician likes Mozart in the morning, he either likes whisky at night or has to wait 20 minutes for a bus; conversely, if he likes whisky at night and has to wait 20 minutes for a bus, he is a mathematician—if he likes Mozart in the morning.

When must a mathematician wait 20 minutes for a bus?

The reader is not advised to try to work out the solution, for this problem is practically impossible to handle verbally.

Although these particular brainteasers are artificial and trivial, in form they are quite typical of problems that arise every day in modern engineering and business operations. Many of the problems are so complex that they cannot be solved by the conventional processes of verbal logic. The necessary facts may all be known, but their interrelationships are so complex that no expert can organize them logically. In other words, the bigness of modern machines, business and government is creating more and more problems in reasoning which are too intricate for the human brain to analyze with words alone.

As a result a number of corporations and technicians have recently begun to take an active interest in the discipline known as symbolic logic. This invention, devised by mathematicians, is simply an attempt to use symbols to represent ideas and methods of handling them, just as symbols are employed to solve problems in mathematics. With the shorthand of symbolic logic it becomes possible to deal with such complex problems as the Pitts conundrum about the mathematician waiting for the bus.

Formal logic, as every schoolboy knows, began with the syllogisms of Aristotle, the most famous of which is: "All men are mortal; all heroes are men; therefore all heroes are mortal." The Greek philosopher set forth 14 such syllogisms and believed that they summed up most of the operations of reasoning. Medieval theologians added 5 syllogisms to Aristotle's 14. For hundreds of years these 19 syllogisms were the foundation of the teaching of logic.

Not until the 19th century did anyone successfully apply symbols and algebra to logic, in place of the verbalisms of Aristotle and his followers. In 1847 an English schoolteacher and mathematician named George Boole published a pamphlet called *The Mathematical Analysis of Logic—Being an Essay Towards a Calculus of Deductive Reasoning*. In it he stated a set of axioms from which more complex statements could be deduced. The statements were in algebraic terms, with symbols such as x and y representing classes of objects or ideas, and the deductions were arrived at by algebraic operations. Thus Boole became the inventor of symbolic logic. His work was followed up by mathematicians in many countries. Their chief aim was to use symbolic logic to solve logical paradoxes and other fundamental problems of mathematical thinking. By 1913 Alfred North Whitehead and Bertrand Russell, using a system of symbols invented by the Italian mathematician Giuseppe Peano, had developed a formal "mathematical logic," which they presented in their *Principia Mathematica* (see "Mathematics," by Sir Edmund Whittaker; *Scientific American*, September, 1950).

Today symbolic logic is an important branch of mathematics, occupying the full time of about 200 mathematicians in the U. S. alone. But the main subject of this article is its practical applications in engineering and business.

Let us first take a few simple illustrations to indicate some of the basic symbols and operations employed in symbolic logic. Any single proposition, however simple or complex, is represented by a letter of the alphabet. For example, the letter a can stand for the statement "The sun is shining," or for something more involved, like

"The three-power commission has been directed to look into the question of whether or not a West German federal police force should be created." Then certain special symbols are used to show relations between propositions. A dot, for example, stands for the word "and." Thus the two-proposition statement "The sun is shining and it is Thursday" can be represented by the expression $a \cdot b$.

The symbol ⊃ stands for the logical relationship "if... then." Thus the assertion "If you love cats, then you are a true American" can be written $a \supset b$. Now by the use of other symbols and by operations similar to those in ordinary algebra, this statement can be transformed into a fully equivalent expression in another form. For example, using the symbol v, which stands for the word "or," and a superposed bar, representing the negative, the expression becomes $\bar{a} \, v \, b$, meaning "You do not love cats or you are a true American." The statement can also be transformed into one containing the symbol for "and." Thus $\overline{\bar{a} \cdot b}$ means "It is not the case both that you do not love cats and that you are a true American," or in ordinary English: "You cannot be indifferent or hostile to cats and also be a true American."

It is important to bear in mind that the symbols have nothing to do with the truth or falsity of the propositions themselves, just as algebra is not concerned with whether its symbols stand for apples or hours. The operations of symbolic logic can only show that, given certain premises, certain conclusions are valid and other are invalid. In this case, assuming that only cat-lovers are true Americans, if you are not a cat-lover the only logically valid conclusion is that you are not a true American, however debatable the proposition may be as a moral principle. The establishment of factually accurate premises is outside the province of logic; its concern is with the validity of the conclusions drawn from a given set of facts or assumptions.

By means of simple signs such as those here illustrated, symbolic logic reduces complex logical problems to manageable proportions. The symbols, like the schoolboy's algebra signs, do much of the logician's thinking for him. Large numbers of propositions can be related to one another in easy algebraic terms; equations can be arranged and rearranged, simplified and expanded, and the results, upon retranslation into English, can reveal new forms of statements that are equivalent to the original or can disclose inconsistencies.

The first application of symbolic logic to a business problem was made in 1936 by the mathematician Edmund C. Berkeley, who is also the designer of the small mechanical brain known as Simple Simon (*Scientific American*, November, 1950). Berkeley, then with the Prudential Life Insurance Company, applied symbolic logic to a difficult problem having to do with the rearrangement of premium payments by policyholders. Every year hundreds of thousands of persons request changes in the schedule of payments on their policies, and there is a bewildering array of factors that must be taken into account in making such changes. The company had devised two sets of rules, intended to take care of all possible cases. Were the two rules equivalent? Berkeley suspected that they were not; that there might be cases in which one rule would call for one method of rearranging the payments and the other for a different method.

His problem was to prove that such cases existed. It was hopeless to try to analyze the possibilities by ordinary verbal logic. One part of one of the rules, for example, stipulated that if a policy-holder was making premium payments several times a year, with one of the payments falling due on the policy anniversary, and if he requested that the schedule be changed to one annual payment on the policy anniversary, and if he was paid up to a date which was not an anniversary, and if he made this request more than two months after the issue date, and if his request also came within two months after a policy anniversary—then a certain action should be taken. These five ifs alone can occur in 32 combinations, and there were many other factors involved.

Berkeley decided to reduce the many clauses and possible combinations and actions to the algebraic shorthand of symbolic logic. The stipulation detailed above, for example, could be written $a \cdot b \cdot \bar{c} \cdot d \cdot e \supset C$, meaning that if the conditions a, b, $\bar{c}$, d and e existed, then the action C was called for. By an algebraic analysis Berkeley was able to show that there were four types of cases in which the two rules would indeed conflict, and an examination of the company's files revealed that such cases actually existed. The upshot of Berkeley's work was that the two rules were combined into one simpler and consistent rule.

Symbolic logic has since been used in many other insurance problems. Mathematicians at Equitable, Metropolitan, Aetna and other companies have applied it to the analysis of war clauses and employee eligibility under group contracts. And other corporations have found symbolic logic very helpful in analyzing their contracts. Contracts between large corporations may run into many pages of fine print packed with stipulations, contingencies and a maze of ifs, ands and buts. Are the clauses worded as simply as they might be? Are there loopholes or inconsistencies? A symbolic analysis can readily answer such questions, and lawyers have begun to call on mathematicians to go over their contracts.

Another interesting use of the technique is in checking the accuracy of censuses and of polling reports. If a public opinion poll-taker reports that he has interviewed 100 persons, of whom 70 were white, 10 were women and 5 were Negro men, it is easy enough to see that something is wrong with his figures. But take an actual case such as this: A census of 1,000 cotton-mill employees listed 525 Negroes, 312 males, 470 married persons, 42 Negro males, 147 married Negroes, 86 married males, 25 married Negro males. Are these numbers consistent? Symbolic logic can give the answer quickly.

In engineering symbolic logic is particularly useful for the analysis of electric circuits. A circuit can be likened to a contract—it has alternatives, contingencies and possible loopholes, the chief difference being that it uses patterns of switches instead of words and clauses.

More than a dozen years ago Claude E. Shannon, then still a student at the Massachusetts Institute of Technology, began to explore the application of symbolic logic to such problems. At the Bell Telephone Laboratories he has recently completed an elaborate analysis of switching circuits by "engineering logic."

Suppose, to take a simple example, the problem is to simplify the six-switch circuit schematized in the left-hand drawing on page 167. The switches are given various symbols. The one labeled C is independent of all the others. The two W switches are connected so that they open and close together. The two S switches also operate together. The sixth switch is designated $\bar{S}$ (not-S, or the opposite of S), because it is open when the S switches are closed and *vice versa*.

There are four possible paths across this circuit from one side to the other. Current will flow across it when the upper S switch is closed, when the C and the upper W switches are closed, when the lower S switch is closed and when the lower W and $\bar{S}$ switches are closed. In the language of symbolic logic this sentence becomes $S \, v \, W \cdot C \, v \, S \, v \, \bar{S} \, W$, with the symbol v, as we have seen, meaning "or." It is at once evident that we can drop one S, since S is equivalent to the expression $S \, v \, S$. The statement now becomes $W \cdot C \, v \, S \, v \, \bar{S} \cdot W$. Next, we can simplify further by dropping the $\bar{S}$, for $S \, v \, W$ is the logical equivalent of $S \, v \, \bar{S} \cdot W$—just as the statement "Williams struck out or Williams did not strike out and walked" is the same as "Williams struck out or walked." This reduces the circuit to $S \, v \, W \, v \, W \cdot C$. A further analysis shows that $W \, v \, W \cdot C$ is equivalent to W. Logically speaking, the statement "Williams walked or Williams walked and was left at first base" provides only one unequivocal piece of information, namely that Williams walked. So the entire circuit boils down to $S \, v \, W$. It can be redesigned in a simple form, illustrated in the right-hand drawing on page 167, which eliminates four "redundant" switches and is fully equivalent to the original.

To use symbolic logic on a problem as simple as this would be like killing a mouse with an elephant gun. But in designing more complex circuits the method may save considerable time and money. At the Bell Laboratories, for example, a group of engineers some time ago undertook to design a special coding instrument. Applying conventional methods of analysis, they produced a 65-contact circuit for the job after several days of work. Then an engineer trained in symbolic logic, starting from scratch without seeing their design, designed an equally successful circuit, with 18 fewer contacts, in only three hours. Today more than 50 Bell engineers use symbolic logic in their work. The method has been applied successfully to a wide variety of problems, but it is not the final answer to all circuit difficulties. Its use is limited mainly to telephone equipment with about nine two-contact relays, which may be in 512 possible positions. In its present infant state even this powerful method of analysis cannot handle the breath-taking complexity of large central exchange stations where a single telephone call may cause the opening and closing of 10,000 contacts.

Perhaps the chief use of symbolic logic is in the design of large-scale electronic calculating machines. Eniac, the first of these machines, contains about 20,000 tubes and 500,000 soldered connections. One of the most important problems in the attempts to build more efficient and more elaborate computers is to reduce the number of tubes, and symbolic logic has been helpful in simplifying the circuits. For example, in building the Mark III all-electronic computer at the Naval Proving Ground in Dahlgren, Va., the engineers decided that a nine-tube circuit was about the minimum that would serve for its adding units. But Theodore Kalin and William Burkhart of the Harvard Computation Laboratory, applying symbolic logic, reduced it to six tubes.

These applications merely suggest the fruitful future that lies ahead for symbolic logic, not only in business and engineering but in science. Wherever complex problems in logical analysis arise, the new shorthand may help to find solutions. One such field is biology, which is beset with a host of complex logical problems. Already Walter Pitts and Warren McCulloch of the University of Illinois Medical School have begun to employ the symbolic logic of the *Principia Mathematica* in an effort to analyze some of the relationships among the 10 billion nerve cells in the human brain. Norbert Wiener of M.I.T. emphasizes that the new study of cybernetics, which analyzes similarities between the brain and computing machines, leans heavily on modern logic.

Although its applications are steadily widening, the major part of the work being done in symbolic logic is still in the field of mathematics. In mathematics this new tool has had so powerful an influence during the past four decades that today some consider mathematics to be only a branch of logic. Mathematicians are applying symbolic logic to examine some of the basic assumptions upon which mathematical theories have been built—assumptions that have long been taken for granted as "obvious" but have never been subjected to rigorous analysis. They are using it to try to resolve verbal paradoxes, which have always baffled logicians: e.g., "All rules have exceptions," a rule which denies itself, since by its own assertion this statement must also have exceptions and therefore cannot be true. Many other basic problems in logic and mathematics are being explored by the new analysis.

Indeed, modern logicians, assisted by the powerful new technique, have punched the classical Aristotelian system of logic full of holes. Of the 19 syllogisms stated by Aristotle and his medieval followers, four are now rejected, and the rest can be reduced to five theorems. Modern logic has abandoned one of Aristotle's most basic principles: the law of the excluded middle, meaning that a statement must be either true or false. In the new system a statement may have three values: true, false or indeterminate. A close analogy to this system in the legal field is the Scottish trial law, which allows three verdicts—guilty, not guilty or "not proven."

Because the use of symbols sometimes makes it possible to determine by purely routine operations whether or not a particular statement follows from given assumptions, symbolic logicians have experimented in designing logical machines. Kalin and Burkhart have, for example, built one that can check Aristotelian syllogisms or solve certain insurance problems, and workers at the University of Manchester in England are developing a more elaborate machine.

Not even symbolic logic will ever produce a machine that can do all man's thinking for him. But some logicians believe that symbolic logic may lead to the construction of synthetic languages that will help to free scientific thinking from the murky tyranny of words.

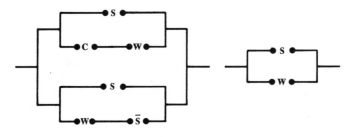

Figure 5-1

Questions for Symbolic Logic

1. What is the purpose of symbolic logic?
2. Who was the first to apply successfully algebraic methods to logic?
3. What are some advantages of symbolic logic over traditional logic?
4. Translate the Carroll syllogism into symbolic logic and attempt the same deduction.
5. Read the article "Simple Simon" indicated in this selection.

Part Six

The Image of Computation

Introduction

Many mathematicians would question the propriety of including a section on computers in a book about mathematics, mathematicians, and what they do. (cf. Halmos' article in this volume.) They would fear that such a section would only foster the common misconception that mathematicians are invariably associated with computers in some way, or the somewhat less common delusion that mathematics and computation are the same. I hope that the rest of the book has helped to expose and eliminate these myths.

Computation, on the other hand, is an important discipline of its own. It forms an important branch of what has become known as the mathematical sciences. As such, it has a considerable amount to contribute. Moreover, a significant amount of mathematics goes into the computing field.

This section will introduce you to some of the devices man has used through the ages to aid him in the laborious task of computation. In addition, I wanted to show you that the men who are most concerned with the manufacture and use of the modern electronic digital computer are among those who are most concerned with its implications.

From: IBM Instructors' Guide

A Brief History of Computing Devices

Calculating machines have a history that is almost as old as man himself. In point of fact, the combination of instrument building and rational mathematical thinking represents a continuous thread that runs through the whole of recorded history, linking together and dominating the pattern of scientific thought and technological development.

While the draftsman and thinker are often placed at opposite poles, they have often combined to make significant contributions to the world. The most direct product of this fruitful union has been the calculating machine.

Prehistoric and Primitive Man and the Earliest Historical Records

If we examine the artifacts of prehistoric man, such as tools, pots, jewelry, and the surviving foundations of cities, it becomes evident that even at this early age, there were fine craftsmen and a deep tradition of handiwork. The art of writing, making its appearance around 3000 B.C. in the ancient river-valley civilizations of Egypt and Mesopotamia, was peculiarly linked with mathematics. Indeed, the

---

Reprinted by permission from C20–1695–*Special Academic Computer Concepts Course for the Humanities, Libraries and Museums.* © 1968 by International Business Machines Corporation.

earliest pieces of writing that have come down to us are accounts listing various numbers of objects. They specify so many men, women and children; so many jars of wine, beer, and oil; this number of pitchers and plates; that number of bricks and planks; that volume of earth to be dug; and so on. These "shopping lists" of the temples and the official treasuries are so common that it almost seems as if writing was invented to aid the keeping of such records rather than to record man's thoughts and literature. By the time these records were preserved, man had already long practiced the art of keeping accounts, of adding and subtracting, and operating with numbers.

In nearly all pre-literate cultures—as today—there existed methods for recording numbers of things and keeping accounts of them. Notches cut on a stick or a bone have been used from the earliest times to represent quantities of objects. The modern word *calculate* comes from the diminutive form of the Latin *calx* (a stone), and refers to the ancient practice of using pebbles as counters. Similarly, seashore peoples used little shells for the same purpose, and all over the world man has used his fingers as an aid to calculation.

However, more complicated ways of recording numbers soon developed, based upon man's skill as an artificer. Jewelry, that ancient art of adornment, was pressed into use in some communities. For example, in Africa the Masai tribe used decorative necklets and armlets to record each passing year in the life of the women. In Peru the *Quipu* was used to handle numbers, and consisted of a row of strings gathered at one end along a retaining strand. Each string bore knots of various degrees of complexity. A single ordinary knot stood for a unit, a knot of double convolution for ten, etc. Thus, at the very onset of recorded history, man was already using objects, both natural and artificial, as an aid to his already developed ability to count.

The First Great Civilizations

By 1500 B.C. the art of writing, already firmly established in Egypt and Mesopotamia, was beginning to extend into the early Mediterranean cultures. The rise of commerce and the need for accounting brought forth a new class of scribal craftsmen who took care of the utilitarian functions of commercial arithmetic and everyday measurement. While the Babylonians seemed to have reveled in the pursuit of mathematics for its own sake, the Egyptians were more practical. Indeed, both the beginnings of mathematical astronomy and the tradition of a mathematical interpretation of nature may be traced to the Old Babylonian period. Similarly, the

Egyptian practice of surveying and measuring the lands and fields vacated by the receding flood waters of the Nile have given us the word *geometry*.

More important in the history of computers, it was in Egypt that the earliest astronomical instruments of any complexity were first developed—namely, sundials and water clocks. However, these devices only incidentally told the time of day, but were rather astronomical peep-shows built primarily for edification, since many of these ancient specimens were covered with astronomical inscriptions. Their role was indeed more scientific than secular. It has been suggested that the inexorable and immutable law governing the daily rotation of the heavens and the movements of the planets inspired man to duplicate them. In this way he was demonstrating both his understanding of the universe and his skill in constructing a replica thereof.

Greco-Roman Civilization

The union of Babylonian numerical analysis with the Greek desire for a tangible and visible model of the universe gave man his first huge success in understanding the world in terms of mathematical reason. This new-found comprehension was employed both to predict the positions of heavenly bodies and to remove the capriciousness that more primitive men had attributed to the whims of the Gods. Thus, the Greeks utilized Babylonian astronomy and mathematics in constructing their own scientific model of the universe, creating in this way a union of arithmetic and geometry.

The exceptional development in the use of mathematics and science stimulated new and important advances in mathematical hardware. Taking the form of models of this newly understood universe, they showed the heavens revolving in the natural cycles, repeating themselves with appropriate rhythms. Alongside these revolving celestial globes was a second category of computing sundials. The sundial was not only a device for telling the time (most of them had hour lines inscribed on their surfaces), but in addition had parts representing the equator, the tropics, the solstices, and the equinoxes. Thus, the sundials embodied the use of geometry to model the basic factors of astronomy.

In addition to sundials, which grew in complexity as many different dials were employed, globes and spheres of all sorts were developed, as well as water clocks of considerable sophistication. It is thought that gear wheels appeared around the time of Archimedes (287–212 B.C.). These examples of mathematical hardware were the first sophisticated scientific instruments exhibiting any real complexity of construction, and may be considered as the first true analog computing machines.

From the very few technical handbooks, like those of Heron of Alexandria and Vitruvius, it is seen that an endless variety of semiscientific mechanisms of considerable ingenuity existed at this time. The texts described the use of gears in simple reduction trains, used, for example, as an odometer that counts the revolutions made by a carriage wheel in order to measure the distance traveled. These writings also described automata worked by strings and levers, as well as complicated theatrical effects produced by water and air pressure, such as dancing figures, singing birds that flap their wings, and temple doors that open by the lighting of a ceremonial fire.

By the time of Ptolemy, astronomy had progressed to the point where it could accurately depict the known universe in elegant and comprehensive fashion. Then as now the problem arose of representing three-dimensional concepts in terms of two dimensions. As a result, special methods were devised for drawing astronomical diagrams on a plane instead of a sphere. Thus began a series of techniques now known as graphical computation. And no sooner had such methods been introduced than they began to be mechanized.

Two important ancient instruments were constructed to facilitate and exhibit this elegant means of mapping the universe. In one, the astrolabe, a series of flat disks engraved with stereographic projections and their scales enabled the astronomer to make his calculations graphically. In the second, the anaphoric clock, the main disk of the star map was made to turn automatically by means of a water clock. The map, seen through a window whose shape corresponded to that of the horizon, gave an impressive public display of the theory of astronomy and the skill of the artificer. By means of both the astrolabe and anaphoric clock, one could know the positions of the stars and their risings and settings, as well as tell the position of the sun at night and the stars and moon by day. While the astrolabe was to the ancient astronomer what a slide rule is to a modern engineer, the anaphoric clock was to them the mightiest and most impressive public demonstration of contemporary science.

A second strand interwoven with these astronomical marvels and gear wheels was commercial arithmetic and calculation. The development of aids to commercial arithmetic took a different line in antiquity from that associated with scientific calculation. The keeping of accounts did not require ingenious mathematical construction and mechanical devices, and indeed required but little elaboration of the primitive method of laying out pebbles and shells. Chief improvements lay in the provision of a specially marked table to keep the piles of pebbles in order, and in the making of a portable miniature table that could be held in the hand that kept the pebbles fitted into grooves so that they could not get lost. The special table, or counting board, was comprised of a set of symbols for units, tens, hundreds, thousands, and so forth; and below them a series of parallel lines on which to set out rows of counters corresponding to various amounts of money that were to be added. The portable form, now known as the abacus, consisted of a small sheet of metal in which were cut a number of pairs of slots: a small slot above with a single bead sliding in it and a longer slot below with four single beads. Each of the four beads represented a unit, and the single bead a five, so that, for example, when the single bead was raised and two of the four were also raised, that column represented seven. Such columns were allocated for units, tens, hundreds, and so on up to millions, the appropriate symbol being engraved in the strip between the upper and lower slots.

Calculating Devices in the Middle Ages

After the fall of Rome, Greek science passed to their inheritors, spreading to such far-flung parts of the world as India, China, and Byzantium, and eventually to the culture of Islam, where it took root, grew again, and then was handed on to medieval Europe. In India, for example, the earliest astronomical texts contained references to animated astronomical models that turned by themselves so as to duplicate the eternal and perpetual motion of the heavens. Similar cases of direct transmission and parallel development occur in tracing the complicated historical development of the numerical systems of the world, the abacus, and other devices of commercial arithmetic. The Chinese *suan-pan* may well be as ancient as the Roman abacus, and the Japanese *soroban*, obviously derived from the Chinese instrument it resembled.

In Western Europe, after the rise of trade and commerce during the late Middle Ages, there are many pictures of the special counting tables, or shop counters, of merchants and moneylenders. The top was divided by inscribed lines headed with symbols of units, tens, hundreds, thousands, and so forth, and often for sub-units of the currency such as shillings and pence for the pound. Special counters (*Rechenpfennigs*) were minted for use with these tables, and the art of casting accounts became the subject of many treatises and manuscripts. These counting tables remained in use until the early years of the sixteenth century, when they began to be displaced by the comparatively new art of keeping accounts by means of numerals written on paper, and sums worked in the arithmetical techniques that came with the consistent use of modern Hindu-Arabic numerals.

However, it was in the area of astronomical instruments and *protoclocks* that calculating devices progressed most during the Middle Ages. In Islam, Moslem scientists treated astronomy not as an abstract mathematical theory, but rather as a physical description of the cosmos to be visualized as a model and measured with instruments. The astrolabe became especially popular, and by the eighth century there existed whole dynasties of special craftsmen who made this and similar instruments, improving them with mathematics and artistic embellishments until they became scientific jewels, highly prized by scholars and their princely patrons alike. Often these astrolabes, as well as other astronomical instruments, were made by a whole team of craftsmen: a metalworker, a mathematician, an engraver and decorator. This pattern of collaborative work is reflected in the very first astronomical observatories, fitted with a great range of instruments and calculating devices and staffed by astronomers, mathematicians, instrument-makers, clerks, and students.

Although the astrolabe seems to have been the most ingenious instrument in general use, there was also a large range of sundials, quadrants, globes, and angle-measuring instruments for making observations. Many of these devices were doubtless derived from earlier Greek prototypes, but in all of them the Moslems made general improvements and adaptations. The most important developments were special analog computing devices for calculating the places of the sun and moon in the Zodiac, indicating the phases of the moon, the age of the lunar month, and following the motions of the planets.

By the thirteenth century, Islamic science had refined and developed the astronomical clock, or protoclock, into a mechanical peep show. There were water clocks with moving peacocks, monkeys, and elephants, systems for ringing the hours on gongs and bells, indicating them by pointers or by doors that open to reveal a mannikin, as well as anaphoric clocks that had turning astrolabe-type dials to show the planets by night and day. One explanation for these astronomical devices was the importance of telling time in the religious practices of Islam, in which the observance of ritual prayers at exact moments designated throughout the day was of cardinal importance.

In Gothic Europe, the great astronomical cathedral clocks represented the high point in the development of scientific machinery. Drawing on the mechanical escapement, which replaced the water clock with a more precise and reliable weight-driven equivalent, the mechanical clock reached maturity with extraordinary rapidity. The earliest example known in great detail was constructed by Giovanni de Dondi in Padua in 1364. It consisted of an array of seven main planetary dials and complicated assemblies of circular and elliptical gear and link mechanisms. De Dondi's clock was followed by many other such devices, indicating a well established tradition of fine instrument-making and skill in the design and production of this special sort of scientific machinery.

Renaissance Instruments and Computing Devices

Two great forces dominated the renaissance of science: the printed book and the rise of scientific practitioners. The former provided new access to the scientific texts of antiquity and brought them before a larger and more diverse audience than there had been for the precious manuscripts within the university and monastic institutions. The logical outcome of this increased spread of learning was the rise of a stable and massive craft industry for the manufacture of scientific instruments. Both Augsburg and Nuremberg, for example, became the chief centers for scientific craftsmanship. Dynasties of fine metalworkers made astrolabes and compasses, surveying instruments, clocks and watches. To them came the professors and mathematicians, the princes of Europe, and the dealers in books and manuscripts.

Later these craft guilds began to establish themselves in other centers. In England, the redistribution of monastic lands under Henry VIII, and the golden age of maritime exploration during Elizabeth's reign ushered in a period when the surveyors and navigators were much in demand, and the country needed computational devices for astronomers, surveyors, and navigators, and gunners. Great ingenuity emerged in the design of such instruments as a cannon-level with a built-in set of scales that enabled one to compute the range of balls of different weight and caliber; a taximeter with a paper tape moving steadily forward as the wheels of the carriage turned; and a compass needle which pressed into the paper to indicate the direction of the carriage motion at every few turns of the wheel.

Calculating Instruments in the Scientific Revolution

By the dawn of the seventeenth century, the art of making technical computations for dialing, gauging, and gunnery became so important and widespread that a number of instrument-makers in several countries began fabricating special ruled scales that were designed to make such measurements and calculations easier. For example, around 1580, Humphrey Cole in London made several such scales, including two folding rules on which were engraved all the calibrations needed by a master gunner. Some ten years later, the young Galileo designed a geometric and military compass which became the most widely used scientific computing device until it was replaced by the slide rule around 1800.

Early in the seventeenth century, another powerful aid to computation, the logarithm, was introduced. By 1617, when they had already begun to revolutionize astronomical calculations, John Napier invented yet another popular aid to computation, the little rods now known as "Napier's bones" each bearing a multiplication table for some particular digit. With a stock of such rods, one could pick those corresponding to give a number of many digits, lay them side by side, and read the result of multiplying that number of any other digit. In this way, multiplications could be performed in an almost automatic fashion. Five years later, William Oughtred, another mathematical practitioner and teacher, brought out a *Circle of Proportion*, a sort of proto-slide. This type of instrument was reinvented and improved many times during the seventeenth century. Special scales of all sorts were added, and calibrating instruments were produced for astronomy and navigation, to determine the weight of metals, and the reactions of chemistry, and for almost every conceivable use.

Along with this activity in the art of scientific calculations came a renewed interest in commercial arithmetic. One of the first steps towards its mechanization is found in William Pratt's *Arithmeticall Jewell*. The book, published in 1677, had bound into its cover a little ivory table with brass sliders. When these were pushed around with a stylus, operations in arithmetic could be performed without writing numerals.

The coming of true mechanical calculation took place around the middle of the seventeenth century, the heyday of the Scientific Revolution. This was when the first great academies of science were founded, when important mathematical techniques were being developed, and new types of scientific instruments were giving man access to observation and experiments in a glorious and inspiring profusion. At this point, a crucial step was taken, bringing together the lines of scientific calculation and its clockwork on the one hand and the ciphering process and its counting boards of the commercial world on the other. The first step seems to have been the recently discovered and reconstructed calculator made in 1623 by Wilhelm Schikard, a professor of mathematics in Tübingen. Characteristically, he called it a Rechenuhr, a "calculating clock." Although its operation was unsatisfactory, it deserves first place as a true digital computer, theoretically able to multiply two numbers by purely mechanical means, using a mechanism of rods and gears and an automatic carrying mechanism for moving tens to the next-highest column.

The next and most famous step was taken by Blaise Pascal (1623–1662), a mathematician, scientist, philosopher, theologian, and writer. In 1642, he devised a completely automatic adding machine that worked by means of a series of toothed wheels, on which the given numbers could be set in turn, communicating their total by register wheels linked to them. The wheels could be turned in either direction so that both additions and subtractions could be handled. It is said that Pascal originally designed and made the machine to help his father keep accounts in his tax office at Rouen. Five years later, Pascal obtained the patent rights for his device, the principle of which was incorporated and adapted into many other similar machines. Described completely in Diderot's *Encyclopédie* (1751), it has traditionally been regarded as the first of the new line of geared digital computers.

A series of three calculating machines of similar nature but developed independently was invented in the 1660's by Sir Samuel Morland, Master of Mechanics to King Charles II. The first of these, invented in 1663 and made a year later by the noted instrument-maker, Henry Sutton, and Samuel Knibb, clockmaker, was an analog calculator for trigonometrical problems. Three years later Morland's adding machine appeared, using little wheels, each furnished with a simple projection, to turn a companion wheel at each revolution so as to carry the tens. Morland's third calculator, brought out in 1673, was a partially successful attempt to achieve mechanical or automatic multiplication.

The basic prototype for a multiplying computer was devised by the mathematician-philosopher Gottfried Wilhelm von Leibnitz (1646–1716). Although conceived in 1671, the first model was not constructed until 1694. Its two essential elements were a collection of pinwheels arranged for adding (similar in principle to Pascal's device), and the new feature of stepped cogwheels, movable so that it could slide to follow decimal places in multiplication. Although the machine of Leibnitz represented the final achievement of a completely automatic arithmetical machine, it was not practical. In spite of the lavish expenditure, the technology of fine instrument-making could not yet reach the precision that was essential, and the calculator remained like those of Pascal and Morland—little more than a pretty toy. It was more than a century before the advanced machine design of the Industrial Revolution could make use of the basic achievement of the ingenious mechanics of the seventeenth century.

The Development of Practical Calculating Machines

The story of eighteenth century computing devices is largely one of experiments in which all possible changes were made on the invention of Pascal, Morland, and Leibnitz, and all the ingenious art of the instrument-maker brought to bear on the ultimate aim of mechanized calculation. For example, the pinwheel was used in the multiplying machine by Giovanni Poleni in 1709, but it proved not much better than that of Leibnitz. In 1775 and 1777, the Earl of Stanhope made two computing machines: the first using the step-gear principle of Leibnitz, and the second using a more complicated variant of the scheme suggested earlier in the century by eminent mechanician Jacob Leupold. Of special significance is the fact that Stanhope's two devices eventually came into the possession of Charles Babbage, who was to continue this tradition and bring it to theoretical perfection. In Germany, the first step toward a multiplying calculator based upon sound mechanical principles was made around 1770 by Mathieu Hahn.

The goal of mechanical reliability was achieved by Charles Xavier Thomas de Colmar (1785–1870). The machine was based upon the Leibnitz step gear which it used in conjunction with a simple system of counting wheels with automatic carrying. The secret of its success was the use of many springs and other contrivances to destroy the momentum of the moving parts so that they would not carry beyond their intended point, a frequent failing of earlier machines. Colmar's *Arithmometer* was the first such device to be put into commercial production and achieve widespread general use.

From this time forward, the history of computing machines divides into two paths: on the one hand there developed a trend towards the perfection of a cheap and reliable machine for commercial use; on the other hand, attempts were made to devise machines capable of reaching further into the complexity of mathematical thought. Thus, during the remainder of the nineteenth century, one group took a step back to the simpler adding machine and eventually made this a common article of office furniture, while the other continued to move the multiplying calculator toward the mass production that would make it a general aid to scientists. At the same time, they bent their efforts towards mathematical machines of different content.

The first monumental efforts to develop a large mechanical digital computer were undertaken by Charles Babbage (1792–1871). Very much a child of the Industrial Revolution, he was obsessed with the idea of harnessing the power of steam to mechanical computation and of typesetting the lengthy mathematical tables used, for example, in navigation and astronomy. That Babbage was able to accomplish anything at all is remarkable in view of the fact that he did so in an age limited to gearwheels and other mechanical linkages, and under a government reluctant to finance such unprofitable research to its conclusion. These two factors eventually left Babbage with two incomplete machines on his hands, embittered and without funds.

His first device, the *Difference Engine*, was able in theory to compute by successive differences and set type automatically so that the output would be in the form of printed tables. The *Analytical Engine* embodying the principle we now know as "programming" depended for its operation upon two sets of punched cards containing nine positions and up to a dozen columns to tell the machine what manipulations to perform at any state of successive calculations. These instructions were maintained in the "Store," an embryonic memory. The set of operation cards programmed the engine to go through a set of additions, subtractions, multiplications, and divisions in a prearranged sequence, while the so-called variable cards stored the actual numbers to be acted upon by these operations. The real computation was done in what was known as the "Mill," a sort of primitive mechanical accumulator. The results were both printed on paper tape and punched on blank cards. Thus, in a very real sense, Babbage's *Analytical Engine* was what we now call a symbol manipulator and a century before the electronic computer possessed both a stored program and a means for encoding information. However, neither one of these concepts was new.

The use of punched holes in cards for the digital storage of information had its origins some fifty years before Babbage in the mechanized weaving of patterns in cloth. Strings of these prepunched cards were used to control the proper lifting in various combinations of the hundreds and hundreds of silk warp threads of the draw loom. In 1801, J. M. Jacquard, capitalizing on principles previously established, developed the first fully successful automatic loom. It utilized a series of punched cards—a complex pattern required thousands of them—held together like links of chain, that programmed the warp and thus the overall pattern of the tapestry being woven.

Returning to Babbage's unsuccessful efforts, in 1834 a copy of one of his scientific papers came into the hands of the Swede, George Scheutz, and his son. After several trial models, by 1853 they had a successful version of a calculating machine that was exhibited at the Paris World's Fair in 1855 and found its way to the Dudley Observatory in Albany, New York.

The *Difference Engines* of Babbage and Scheutz provided a new sort of automation of mathematics. Previous devices had enabled man to mechanize only single operations: addition and subtraction for the Pascal type, multiplication and division for those of Leibnitz variety. With the *Difference Engines* came the new concept of a continuous series of operations, the taking of differences that could be used to build up a printed table produced automatically for almost any regular analytical mathematical function that one wished to tabulate. Some of the techniques introduced by these two inventors contributed significantly toward the later commercial perfection of mechanical computation machines, which became standard equipment for the Scientist's work-bench a century later. However, the grand concept of an automatic and powerful computing machine had to wait for the development of new electronic skills capable of achieving those ends for which mere metal, wheels, and mechanical linkages were too inert, imperfect, and tedious.

From: IBM Instructors' Guide

Introduction

Before the invention of the computer, learning was a property almost exclusively ascribed to the self-conscious living system. Both logic and memory (that is, reasoning) were unique characteristics of the human being, making him superior to all other forms of life. Now, however, a computer can be programmed to play checkers and to "learn" from its past experience and improve its own game. Indeed, in a movie made some time ago, a checker champion pitted his skill against an IBM 7090 computer, programmed by Dr. Arthur Samuel. It was a close match between a man who spent a lifetime learning the game and the computer that had learned the rules in half an hour. Not only that: programmed to look at all possibilities several moves ahead, the computer would have beaten the champion every time! How would a man feel if a machine could surpass him in the very thing in which he takes the most pride?

A second disturbing thought concerns machines that have the capacity to reproduce themselves. Western man has commonly held the belief that God created life, and that only living organisms possess the capacity to reproduce. In his desire to glorify God with respect to man, and man with respect to matter, it was natural to assume that machines cannot make other machines in their own image—that, in other words, a sharp distinction exists between living and nonliving, between creator and creature. But now, as Norbert Wiener has pointed out, machines can be constructed—a transducer, for example—that are able to make other machines, reviving the Jewish legend of the Golem, a robot giving life through cabalistic rites.

On the one hand, Dr. Warren McCullough, an eminent neurologist, mathematician, and engineer, believes that machine might even survive man, carrying on in the same direction, standing on man's shoulders. He sees these devices as purposeful, perhaps even endowed with the ability to feel. On the other hand, Dr. Margaret

Reprinted by permission from C20–1695–*Special Academic Computer Concepts Course for the Humanities, Libraries, and Museums.* © 1968 by International Business Machines Corporation.

Mead, the well known anthropologist, says that this is anthropomorphism, a projection by those who think about machines, either of the fear of losing their own autonomy or of a tremendous desire for power. She sees this as the contemporary cosmology of our age, when man is questioning his own position in the universe. Preoccupied with the state of his technology and the destructive potentialities of scientific knowledge, man transfers this fear onto images of machines taking over and man disappearing. It is a mythological way of talking, a poetic notion of machine robots taking over from man—a new form of God created in man's own image.

Thus, the machine has forced us to alter our notion of superiority, leaving us vague about our own human uniqueness. What, then, is man that a machine is not? In the long list of mounting uncertainties, none has gripped the mind more than this single question. In technology's assault on traditional concepts of the universe and man's place in it, none has given more challenge to our age-old faith in ourselves as somehow especially endowed. All of our religious and philosophies throughout history, even our systems of law and democratic political institutions, postulate the human being as deserving of a mystical pre-eminence among all other phenomena of nature and the universe.

If one examines the prevailing attitudes towards the machine closely, it soon becomes apparent that there are two complete independent currents of thought and emotion. The prevailing viewpoint regards the computer as a beneficial tool of man. However, there is a secondary undercurrent of uneasiness that is related primarily to the notion that the computer is an autonomous thinking machine. It is interesting that the central focus of this suspicion of computers is not the threat of automation and job displacement, but that it is rather the idea that there is some sort of science-fiction machine which can perform the function of human thinking—functions which were previously thought to be the unique province of the human mind. This is construed as a downgrading of humans, and engenders a feeling of inferiority in relation to the abilities of the computer. The anthropomorphic notion of a machine which can possibly out-think man is not easy to assimilate or to live with. It suggests that man is less unique than he thought and that he is therefore less important.

The Newtonian Machine

Perhaps one of the basic difficulties in understanding the computer as a logic machine is that it is singularly unfamiliar to us, different from any machine we know. Its parts, we are told, are furiously busy, yet the eye sees no movement, no work. In the past, logic machines have been devices whose functioning was open to the eye, such as a wheel that turns on an axle moved by consecutive shoves of gear tooth against gear tooth, powered by a horse, a man, or an explosion within a cylinder. Machines, furthermore, are predictable; one pulls a lever and something happens. The long train of gears, for example, produce some foreseeable motion, and when it stops, the motion stops. This mental focus, which almost all of us still share, is the legacy of Sir Isaac Newton to whom the whole physical universe became comprehensible through a single metaphor, that of a giant clockwork machine.

Newton achieved something more than heroic insight and breathtaking mathematical elegance. He bestowed upon a world troubled by the breakdown of the Aristotelian cosmos—a universe of self-evident, axiomatic scientific truths—a new reassurance that it moved in a predictable manner. In other words, it was a rational universe in which if something moved, then something else followed. Until the Computer Age, man supplied the "if"—the control—by manipulating a lever, stepping on an accelerator or a brake. Once this was done, the "then" follows just as night follows day, in a predictable fashion. Little wonder, then, that most of us who are still Newtonian-minded rationalists become troubled by the so-called logic machine. Where is cause and effect in a machine which can alter its own behavior? If it is a logic machine, where is the familiar, reassuring Newtonian kind of machine logic? Is that young operator at its console really exercising control of its work? Or is the machine exercising its own control beyond him? If this is so, isn't it really an unpredictable machine outside the framework of Newtonian determinism?

The great ancestor of our machines was the prehistoric plow drawn by a beast of burden. Here, for the first time, man supplemented his muscle power, and most of the energy he now expended was no longer in doing the actual work, but in exercising control over it. However, work and control were clearly separable. Today, we have supplemented the plow with a vast repertory of machines that are no longer powered by the energy of muscle, but rather by chemical energy: the controlled combustion of coal, gas, petroleum, electric power, or the fiery flux of the nuclear pile. But in all these so-called Newtonian machines, control is still in man's own head and hands. Guided by our brain, our hands and toes determine how fast we drive, where we go, and when we stop.

Of course, before the computer there were myriad numbers of automatic milling machines, looms, typewriters, autopilots, and all the so-called self-regulating devices. But in all of these man merely imprinted his control on a tape or a pointer setting, and through these, guided the machine. Work and control were still clearly separable. It is obvious that the conventional Newtonian machine, with its mechanical limitations, is utterly incapable of altering its own behavior, since this behavior is determined by man. Only the orders from men imprinted in the tapes or needle settings have been varied, and the machines are actually responding to these variations. Therefore, no matter how complex these machines become, they cannot alter their own behavior, which is always predictable and built into their very structure. Neither, quite obviously, can these machines learn to play checkers!

The Logic Machine

How, then, does the traditional Newtonian device differ from the logic machine? First of all, the former worked essentially by clockwork, employing the familiar repertory of gear wheels, levers, and jackwork. Their operation is determined by patterns set up in advance. There is no deviation because the "commands," or instructions, are transmitted to the machine and stop there: it is a one-way type of communication. Modern automatic machines, however, such as the space capsule, the proximity fuse, the electronic door-opener, or the control apparatus for a factory, possess sense organs; that is they perceive messages coming from the outside, such as light falling on a photoelectric cell, for example. In other words, the machine is conditioned by its relation to the external world. Its actions are dependent upon feedback: the results of its behavior are transmitted back to it as the basis of its continued—and alterable—operation.

Secondly, Newtonian machines have undergone a kind of reverse evolution. The more sophisticated they have become, the fewer parts they contain. For example, the automobile engine—even now a fairly primitive kind of machine—contains many more parts than the more modern and advanced jet turbines that are far more efficient. In other words, in machine designer terms, the fewer parts a machine has (such other things as performance being equal) the more elegant the machine. A second point is that the components of even the most sensitive parts of a Newtonian engine are machined to a fidelity of perhaps something in the order of ten-thousandth of an inch, which means that a built-in error of one part in ten-thousand is tolerable (it's even called *tolerance*), and most parts of these machines will even tolerate a far higher proportion of error. We also expect a decently high order of reliability, requiring that our automobiles travel something like 20,000 miles without a major overhaul—in other words, 5,000 to 20,000 miles before the machine tolerances, through wear and tear, exceed permissible limits. This means that we want the machine

From: IBM Instructors' Guide

to operate with moderate reliability through about 50 million revolutions of its wheels, or about 400 hours of trouble-free work.

To summarize, Sir Isaac Newton's legacy to us all was an incredible explosion of technology, but a legacy nevertheless of machines conceived and dedicated to the concept that work and control are separate; that the machine's work is always predictable and determinant; and that the fewer parts the better, each with high fidelity and reliability.

The computer, on the other hand, apparently contradicts these Newtonian laws. Indeed, its basic concept, a structure built out of a single, tiny, logical circuit, implies that the more parts, the more circuits, the more powerful and better the machine. This is certainly not Newtonian. The innards of even a medium-size computer contains some 64,000 elements, and tens of thousands more parts in the supporting equipment. Thus, one could almost say that in the computer the more parts the better. Still another point is that whereas acceptable fidelity in the older machines is measured in tolerance of perhaps one part in ten thousand, only absolute, total, undeviating precision can be tolerated in the logic machine: in other words a tolerance of zero parts in a million or a billion. For, obviously, if anywhere in the circuitry of the computer a single junction reads "yes" when it should read "no"—zero instead of one—the machine will pursue with implacable and relentless logic the wrong answer through all of its millions of circuits. Therefore, in the computer, fidelity must mean perfect in tolerance, and reliability must also be total and absolute. Since no tolerances for error are permissible, none can be permitted to develop as the machine does its work. Remember, too, that it performs not at 50 million operations in 400 hours like an automobile engine, but a couple of million operations a minute.

Thus there are startling differences indeed in the engineering psychologies involved in producing this new kind of machine. In high contrast to the predictability and determinism of the Newtonian device, work and control are clearly indistinguishable in the computer. Whereas the older machine cannot alter its own behavior, the computer can, with its feedback capabilities, alter its own behavior when necessary. It does so, of course, by changing its programming, its basic procedures, according to the feedback information it receives from its own activity. Thus, the logic machine modifies the instructions men have given it, taking over control itself.

This is a fundamental and highly important point. Since the logic machine can alter its own behavior, is it therefore unpredictable and indeterminant? If so, it may be potentially self-determining; while it now seeks the data of a problem and decides on the best way to solve it, it may one day define the problem. In other words, can a logic machine, a computer, *think*—now or ever?

The Two Modes of Perception

The question as to whether the logic machine can, now or in the future, assume some or all of the so-called sublimity of man has increasingly troubled philosophers, clergymen, psychologists, and the natural scientists alike. The philosophical implications of the machine that thinks run so deep that the philosopher Sidney Hook convened a symposium at New York University of some of the leading philosophers, theologians, and scientists on the subject of science and the mind of man.

As one philosopher pointed out, all of us except the psychopath have both *public* and *private* experience, or perception. On the one hand, our public perceptions are those that we share with every other man: those perceptions that the objective scientists can measure with his instruments. Thus, we can measure the intensity of sunlight, the exact predictable force of the recoil of a gun when fired, the acceleration of a falling body. In fact, all the phenomena that we now think of as governed by the Newtonian laws of physics—all measurable phenomena—are in the domain of public perception. On the other hand, each of us also perceive privately. Descartes proclaimed, "I think; therefore, I am." Each of us can say this for we perceive it in ourselves, but we cannot with the same clarity and distinctness know it of anyone else, though there may be some justification for publicly assuming it. Thus, "I am" is purely private knowledge. So, too, is our pain for which no instrument of medical science exists to define its character, quantity, or intensity. It is purely private perception, and there is no way of making it public. Similarly our feelings of love are private, and only the poet, not the scientist, can attempt to describe them.

The philosophers at the symposium also argued that these two modes of perceiving are coexistive. True, a private, subjective perception—a moment of terror, for example—is not necessarily accompanied by a public perception. But on the other hand, a public perception is always accompanied by a private one. Thus, each one of us feels the law of gravitation individually, for only the individual knows that he is seeing or hearing or feeling.

But what precisely is it that we sense? And how do we organize the sensory data or fragments into patterns or wholes? Consider for a moment a series of musical tones. A solitary note is not a tune, nor is the note that depicts it a score. But our human perceptions somehow fuse these isolated tones into a continuity and identity. We perceive the relationships among the tones and can even at the same moment name the music, the composer, the instrument being played, and perhaps an association or two it summons from where we once heard it. Thus something further emerges from those isolated sounds. What the music is to us will vary in each, even though, technically speaking, it is purely an assemblage of waveforms or frequencies. Synthesizing

From: IBM Instructors' Guide

the tones into music, perceiving the whole beyond the parts, grasping the form that we call music is something no machine can do. For while we are actually perceiving in time, note-by-note, one at a time, we are also experiencing the whole in a non-temporal way: We are fusing previous tones—the data of the past moments in time—with the present. In machine terms, we are doing more than the gears or circuitry allow. But while the individual hears only one fragment at a time, he also transcends time, retaining what has just passed and looking forward for more to follow. In other words, he is able to perceive the form of the whole message, fill in what has been left out, and deliver the whole: he perceives *meaning*. The machine cannot do this. It is a helpless slave to time, for it simply has no way of sensing more than one element or datum at a given instant. There is no such thing as form, or meaning, or wholeness to the computer. It senses one hole at a time, "reads" one instruction at a time. True, it may be so with incredible speed, but whatever form may be implicit in the data is quite beyond any power of the machine to perceive.

To the human being, any one of these pieces of data may in itself have a special, private significance or meaning beyond perception—in other words, a symbolic value. A red light means "stop"; a hole in the third column of the second row of the punched card means "two withholding exemptions." The binary number, 1101, is translated into the decimal number, 13. In other words, it is we who supply the meaning, not the machine. The computer senses only the magnetic equivalent of 1101 and that is all. For this number to have further significance to the computer requires that we program the significance we see into the machine so that, depending on the program, the number 1101 instructs it to "go to the thirteenth tape," or "write a check for $13," or "order 13 gross of lollipops," or any other action the machine's programmer assigns to the number 1101. To the computer, however, with no sense of the past or the future, the number 1101 means nothing; it is a meaningless instruction.

Suddenly, in realizing that in its very concept and construction, the computer can conceive no meaning, no form, no whole, no continuity, no past, no future, and no time except this very instant, it is evident that no matter how indistinguishable work and control may be inside it, the machine's function is as predictable as logic itself. What does this mean? It means that the logic machine can function only in the finite, mechanical world of publicly measurable phenomena and perception. It is a Newtonian, as determinant, and as predictable as the plow and the pulley, the gear train, and the automobile. There is no way for it to have private or indeterminant perceptions, for it has no way to perceive unpredictability or irrationality with one-at-a-time vision. It cannot judge: it can only calculate. It cannot synthesize: it can only analyze and decide. It is as purposeless as a jet engine provided that what it is instructed to do is as unerringly logical as its electronic circuitry demands. But beyond that—beyond logic—it is quite literally a useless hunk of metal.

Man and Machine

Therefore, can the machines think? If by "think" is meant only man's ability to reason logically in publicly measurable Newtonian terms, then the answer is yes. The computer is able to manipulate arithmetical and logical formulas, and in rapidly accomplishing these operations it *seems* to be thinking. But in point of fact, rather than working with abstract thoughts it is only switching electric currents along pre-ordained paths. In producing answers to specific questions the computer assumes an almost brainlike quality.

However, if by "think" is included some of those purely private and irrational human qualities—qualities such as perception of pain, love, and beauty, and most important of all, our human ability to create them—then the computer is forever barred from human thought. It has no vision. It is blind except for the infinitesimal moment of time it requires to perform one action. In other words, in many ways the computer can perform like the publicly measurable, neurophysiological organism we call the *brain*, but it has no way of ever simulating the private, subjective, both rational and irrational human *mind*. Although the philosophers, psychiatrists and theologians still disagree as to exactly what the private human mind and spirit may be, they do agree, as do the scientists, that it is something quite distinct from the mere brain and from the logic machine.

Consequently, whereas the mind can act on insufficient and disorganized data—making generalizations on its own or reaching conclusions nobody told it to reach—the computer won't work unless it is instructed in detail concerning every step it must take. In other words, human thinking is so marvelous and mysterious a process that there is really not much hope of imitating it electronically. The human brain is uncanny. It programs itself. It asks questions and tells itself how to answer them. It steps outside itself and looks back inside. It wonders what "thinking" is. No computer ever wondered about anything.

But whatever the mind is, and whatever the computer may be, it is obvious that the two—man and machine—have become inextricably interwoven into a single system, and this is but the very beginning. With explosive rapidity, the computer manipulates its meaningless ones and zeros and yields a changed constellation of binary digits translated back into human words and symbols, meaningful only to men's private minds. Man must put in, and man must take out, for minds alone can perceive form and wholeness. Minds alone know purpose or can see the future. The machine knows nothing. It is an idiot performer of incredible agility and speed. Mind and machine have been interwoven into a system, interlinked, interdependent, and together make astounding human achievements. It would be impossible without the machine, but without the mind, it would be nothing. This mind-machine system is of necessity even closer knit; for since the computer of itself is purposeless, it is helpless to monitor the value of its own progress.